ARTHUR H. ROBINSON

THE LOOK OF MAPS

An Examination of Cartographic Design

ESRI PRESS
REDLANDS, CALIFORNIA

Originally published by
The University of Wisconsin Press
Madison, Wisconsin USA
uwpress.wisc.edu

ESRI Press, 380 New York Street, Redlands, California 92373-8100

First edition 2010
14 13 12 11 10   1 2 3 4 5 6 7 8 9 10

Printed in the United States of America

Library of Congress Cataloging-in-Publication Data
Robinson, Arthur Howard, 1915–
The look of maps : an examination of cartographic design / Arthur H. Robinson.
p. cm.
Originally published: Madison : University of Wisconsin Press, 1952.
Includes bibliographical references.
ISBN 978-1-58948-262-3 (pbk. : alk. paper) 1. Cartography. I. Title.
GA105.R63 2010
526—dc22   2010003604

Ask for ESRI Press titles at your local bookstore or order by calling 800-447-9778, or shop online at www.esri.com/esripress. Outside the United States, contact your local ESRI distributor or shop online at www.eurospanbookstore.com/ESRI.

ESRI Press titles are distributed to the trade by the following:

In North America:
Ingram Publisher Services
Toll-free telephone: 800-648-3104
Toll-free fax: 800-838-1149
E-mail: customerservice@ingrampublisherservices.com

In the United Kingdom, Europe, Middle East and Africa, Asia, and Australia:
Eurospan Group
3 Henrietta Street
London WC2E 8LU
United Kingdom
Telephone: 44(0) 1767 604972
Fax: 44(0) 1767 601640
E-mail: eurospan@turpin-distribution.com

It is just as important to study the proper and effective use of various forms of graphic presentation, as it is to study the values of different methods, treatments, grades, and forms of verbal presentation.

—WILLIAM MORRIS DAVIS

Contents

About this reprinting

ESRI PRESS is honored to republish Arthur H. Robinson's pivotal study of cartography, *The Look of Maps*. Originally published in 1952, the book investigates the discipline of cartography as an important intersection between science and art. To that end, Robinson covers a range of topics related to the visual characteristics of cartographic technique, including lettering, structure, and color. He also offers advice that even the modern cartographer will find relevant: adopt a "healthy questioning attitude" in order to improve and refine the graphic techniques used to present information visually through maps.

For this new edition, the text was edited for clarity and consistency; minor wording and punctuation changes were made only in cases where it was unclear what the text meant. Time references were left as they were originally written, and the bibliography and graphics appear just as they did in the first printing by the University of Wisconsin Press.

ESRI Press is pleased to make Robinson's cartographic classic available to readers once again.

Foreword

ONE of the major aims of education is to impart an appreciation of what and how much we do not know. It is primarily with this thought in mind that these essays are presented.

I am acutely conscious that the reader may be reminded of that unhappy person who tells most of a (supposedly) good story—and then forgets the denouement. For the truth is that the unraveling of many of the mysteries of cartographic design and presentation has not yet been accomplished. Nevertheless, in the hope that the half-told story will excite the curiosity of others to investigate further, these essays are presented without apology, but with the hope that the reader will be understanding enough to maintain a constructive attitude—at least toward the subject matter. There is probably room for argument, however, on the assumption hereinafter made, that the "art" in cartography should be considerably more objective than it has been in the past. It would, indeed, be a pleasant state of cartographic affairs if the profession were staffed with geographers who were also accomplished artists and who, when making a map, could design appropriately for the purpose at hand. I do not think that such is the case. In lieu of it, it seems only reasonable that he who is not so subjectively gifted should try to come closer to the ideal via objective means, so that he may not, "in his attempt to be elegant, . . . succeed only in being ridiculous."

The subject of these essays has long been of interest to me. The Second World War postponed (fortunately) the preparation of my PhD dissertation. Instead, it was my lot to be placed in charge of the Map Division of the Office of Strategic Services from 1941 through 1945. During this period our experience in the Cartographic Section of the Division clearly showed that the creation of a special purpose map was frequently as much a problem in design as it was a problem in substantive compilation. It was also apparent that the artist was better fitted to solve the design problem than was the conventionally trained geographer-cartographer. That experience made such an impression that I proposed the visual aspects of cartography as the new subject for a dissertation. The Department of Geography at Ohio State University approved. Hence, much of the research reflected in these essays was undertaken.

The reader may well be nonplussed by the lack of illustrations in a book on cartographic design. The author faced a dilemma. On the one hand the illustration of "how to do things" implies a textbook, which this is not; on the other, to request permission to reproduce examples of "how *not* to do things" is likely to inhibit cooperation, to say nothing of friendship. A third, and perhaps more important reason is that the scientific special purpose map rarely should be examined out of context, so to speak, for its *raison d'être* determines or limits, to a considerable degree, its visual character. To attempt the illustration of the concepts herein considered would have required so many assumptions that the author is convinced the reader would think the results too artificial.

It is a pleasure to acknowledge the aid and encouragement received from a number of quarters—first, to the Department of Geography at Ohio State University who allowed such a geographically unorthodox research subject, and especially to Professors Guy-Harold Smith and Roderick Peattie. The former was particularly helpful, and without his liberal attitude the research could not have been done at all. My first real experience with practical cartography took place before the war while doing illustrative work for Roderick Peattie. From him I learned the value of the unorthodox. Professors Hoyt Sherman and Robert M. Coffin of the Ohio State University's Department of Fine Arts were helpful and very understanding with one with little formal training in design and the visual principles behind it. A number of persons were generous enough to read all or portions of the manuscript, and their constructive criticism is greatly appreciated. These included John K. Wright, O. M. Miller, and Richard Edes Harrison.

The Social Science Research Council awarded a demobilization grant which enabled me to work full time on the subject for two entire summers while at the University of Wisconsin. That institution not only allowed me to be free during those critical summers of heavy post-war teaching (1946 and 1947), but also made it possible for the summer of 1950 to be devoted to preparing the essays.

Without the encouragement, editorial assistance, and criticism (and patience!) of my wife the writing of these essays would not have been nearly so pleasant.

ARTHUR H. ROBINSON
Professor of Geography

University of Wisconsin
July, 1951

THE LOOK OF MAPS

I

Toward a better cartography

IN THE years to come it is likely that the twentieth century may be designated the golden era of cartography. The field once before (in the sixteenth century) experienced what amounted to a renaissance and reformation from the traditions originally established by the Greeks. The current rapid development may be a second reformation.

In most aspects of the entire field of cartography unprecedented strides have been made. The airplane, photogrammetry, air photo interpretation, sonic sounding, radio, electronics, high-speed reproduction, and a host of other developments (including two world wars) have advanced the field to a stage where it is now almost impossible for one individual to embrace it in its entirety with any degree of competence. Yet the cartographer of the past, the one who designed and drew the actual maps, is still the central character in the long process from the gathering of the data to the printing of the map. His work with lines, colors, lettering, symbols, and the other graphic media is the work that makes the data intelligible to the reader. This is the essential cartographic technique. Without this we do not have real cartography—only engineering, mathematics, and printing.

The cartographer does much more than design in the graphic sense. He selects, generalizes, and researches, but in the end he must put his materials and determinations into graphic form.

It is in this aspect of the profession (if we may call such a broad field by that term) that the smallest strides have been made. The ability to gather and reproduce data has far outstripped our ability to present it. There are, however, indications that efforts and thoughts are being expended in this direction. Most of the graphic design conventions were established prior to the current growing acceptance of functional design as the basis of creative effort. That this philosophy has not yet been generally recognized by cartographers is probably not due to any congenital conservatism in the breed, but rather to their inability to find time for research along the ramified lines of visual perception. The treatment of these essays is mainly aimed toward the small-scale special purpose map, but the principles discussed are applicable in some degree to any scale cartography.

Until very recently cartography has suffered from lack of recognition as an academic subject. Although geographers and geography departments in the universities of the United States have long considered maps the prime tools of their discipline, neither they nor their administrative superiors considered it worthy of more than minimum support. Research in the cartographic technique languished until recent years. The Europeans, especially the English and the Germans, have devoted more effort to the technical aspects. Max Eckert, the great German cartographer, is the only writer of the era to examine exhaustively the bases of cartographic methodology. His concept of the "science of cartography" is somewhat confusing, partly as a result of the fact that he wrote it before the trend toward functional design had progressed far, and partly because much of the investigative work in allied fields had not yet been carried out or was only in its early stages.

Considerable analysis of certain concepts in the technique have been made, both in this country and abroad, but they deal primarily with aspects substantive in nature. For example, studies concerning the representation of slope, population, and other distributional facts are devoted first to methods of deriving the information, and second to techniques of resolving the data into a form that allows it to be presented graphically. The analysis of the presentational strength or weakness of the techniques, when included, has usually been incidental and incomplete; but more commonly it has been ignored completely. Presentation in graphic form is dependent upon the use of basic visual techniques such as lines, shapes, letters, colors, shading, position, and a variety of visual media. An analysis of the value of the substantive techniques cannot be definitive unless there exists also a set of principles derived by evaluating these basic visual techniques. There is no way of determining the effectiveness of capital letters of various sizes as symbols for mineral distribution until the characteristics required of point symbols have been catalogued. The latter cannot be accomplished with any degree of success until the underlying bases have been determined and evaluated without the confusion accompanying their intellectual connotation.

These essays will attempt to review some of the visual characteristics of the cartographic technique in their most elementary aspects. The visual characteristics arrange themselves under three general headings: lettering, structure, and color. Each of these subjects will be discussed for the purpose of analyzing its inherent characteristics as they are manifested in cartography. The technical bases for methodologic choice in cartography are largely dependent, at present, upon the research done in other fields. Although this research is not directly applicable to cartographic procedure it provides the necessary groundwork. Whether the experience of related fields will be utilized and developed in cartography remains to be seen.

The placing of methodology on analytical and experimental bases may seem to suggest that creative interest may thereby be stifled. The use of such bases will not result in stereotyped maps any more than adherence to established principles has resulted in all advertising being in the same pattern. The purposes and uses of maps are so varied and the subject matter so infinite that it is improbable that the field will develop many standard procedures outside the topographic map and chart classes. Certainly there will be some, but if cartographers develop a healthy questioning attitude, any such procedures will necessarily be based on sound principles. Fresh ideas and research together with new methods and combinations of techniques will maintain the essential independence of the cartographer.

II

The cartographic technique

THE story of the development of cartography from its beginnings is essentially a composite of the chronicles of exploration and survey, together with such abstruse material as the mathematics of map projections. Until recently the use of maps was largely limited to specialists such as navigators, surveyors, military planners and the like, and the preparation of their maps constituted a problem only with respect to the accuracy of the things mapped. Except for the commercialized atlas trade, presentation media and techniques have been of relatively little concern throughout the development of cartography.

Only in the past few centuries have really major advances occurred in cartographic technique. The majority of these advances, such as the isoline, the graduated circle, and the hachure, have come about because of the ever-present and fundamental problem of presenting quantitative facts. Accuracy is obviously the first objective of any scientific activity; but when presentations of factual materials become widely used, the manner of presentation becomes of primary significance. This is of extraordinary importance in the field of cartography and approaches critical consequence with respect to the special purpose map, the map which treats but a few categories of data. The navigator and engineer are primarily concerned with spatial data of a precise and numerical nature. For these data various kinds of charts, topographic maps, and the language of mathematics are reasonably adequate means for their display, but the same cannot be said for the social scientist who is commonly concerned with less precisely definable data, relationships, or concepts.

Maps in social science as well as its written terminology deal more with the qualitative and interpretive aspects of investigation and knowledge. The vehicles for presenting such materials should be capable of recreating in the mind of the reader, so far as possible, precisely the intended intellectual meanings and interpretations of the author. It is relatively easy to accomplish this when language is the medium, since there are generally accepted standards and definitions for its use. Words are merely intellectual ideas graphically presented; but, unlike most other visual symbol forms, they pass through the eye to the brain without occasioning major visual stimuli. There are, to be sure, significant variations in the readability of typefaces, point sizes, and page layouts, but long usage has tended to submerge these to such an extent that it is doubtful if any significant erroneous intellectual reactions could be ascribed to them. In cartography, on the other hand, the graphic techniques and media are legion, and the possibilities of arrangement are so tremendous that the inexpert map reader, when viewing qualitative, and even some quantitative data, cannot help but unconsciously receive from the unfamiliar visual forms many sensory impressions in addition to, or in place of, the intellectual concepts intended by the cartographer.

Every map is a complex stimulus, for all its shapes have both visual and intellectual relationships to one another. Anything existing within the neat or trim lines may be described as a series of related intellectual concepts represented by visual media. In some cases the visual symbolism is so characteristic or well known that one is able, unconsciously, to disregard the visual stimuli and see and recognize clearly the intellectual concept. In other cases the medium acts more like a cleverly constructed mask and so obscures the intellectual thought that the only way to determine its identity is through reference to a legend or a key. It is only natural that the eye and the mind frequently make guesses concerning the nature of the things thus visually represented, regardless of the earnest intention on the part of the viewer to accept only known facts. This must happen with remarkable frequency, since it is well known that the mind and the untrained eye do not ordinarily distinguish between intellectual and visual stimuli. Indeed, for many people it is next to impossible, for the major part of knowledge comes to all of us through our eyes, and it is only natural to confuse visual and intellectual factors. "Seeing is believing!"

In the recent stages of cartography there seems to have evolved a well developed and unusually tenacious habit of substituting convention for technical quality. It appears somewhat anomalous for conventions to have a strong hold over a field which prides itself on its creative artistry. Nevertheless, from colors to boundary lines, and from lettering to projections, the field of cartography leans, or rather reclines, on its conventions. The habit is insidious for it tends to stifle investigation toward greater technical competence, because the innovator is reluctant to depart from the rut of convention. Even standardization among small-scale maps has its proponents! Many of the standards suggested will no doubt turn out to be ill-conceived conventions. As one distinguished observer, O. M. Miller, has put it, "once a map convention has been established it is difficult not to feel prejudiced in its favor."[1] Until the past century or so, this powerful influence was sufficient, considering the technical limitations of the field, to maintain the profession in good standing with the other arts. As the technical horizon widened, however, convention only gradually relinquished its hold on cartography. Notwithstanding, there were many changes (the cartography of 1910 is very different from that of 1810) yet the power of convention had not greatly decreased. The force simply transferred itself to other technical phases. The advances of the past one hundred years have been enormous but, generally speaking, convention has merely replaced convention.

Examples are many. Perhaps the most widely publicized in recent years is the general use of the Mercator projection. This projection, designed for a specific purpose, navigation, has been regularly and indiscriminately used although many other projections more suitable for general purposes have long been known. Only because of our recent so-called "entrance into the air age," and the vituperation against the Mercator accompanying the move, has this projection declined in popularity. It is interesting to note that, as in the past, the conventionalism, rather than disappearing, seems to be in the process of being transferred, this time to the polar aspects of the azimuthal projections.

Many conventions are logical or stand functional analysis. In this category may be placed many of the techniques of symbolism such as dots, circles, or squares for cities; the pictorial kinds of symbols such as hatched lines for railroads; dot-dash lines for political boundaries; and the innumerable *conventional signs* on topographic maps. They constitute a kind of cartographic shorthand and, so long as they are employed with proper regard for their effect on the overall design of the map, they are adequate for indicating character and location. The importance of the visual functional evaluation of spatial symbols is well illustrated by J. K. Wright who points out that although "tiny men, or ears of corn, or cows . . . may be in better 'harmony' with the things they represent than flat colors or shading would be, they may also be out of harmony with the purpose of the map if that purpose is to give a clear and clean-cut concept."[2]

Some conventions are justified on the grounds that they have been tested by time and found good. All too often, however, they have been tested only by their makers, and the quality of parental pride is not always objective. Basic analysis of their visual effects and logic is conspicuous by its absence. The prime example of such a convention in cartography is the practice of presenting hypsometric layers somewhat according to the spectrum, that is, with green for the lower altitudes ranging upward through the yellows and reds. This system of progression has even been given a kind of international approval, for it was chosen as the basis of the system of representing hypsometric data on the international map. Its champions point to the fact that most people are familiar with the system and thus it "has been found by experiment and experience to give a graphic visual impression of relative altitudes." Similar conservative rationalization opposes most proposals of change. When analyzed, the only justification for this convention is its conventionalism. The spectrum bears not the slightest relation to altitude. The ocean is only occasionally blue; lowlands are not universally green; and mountains are not red. Even more important is the visual fact that the progressive colors of the spectrum bear little relation to one another as far as the eye is concerned. Actually their lack of relation is much more significant in vision. The wide variation in brightness and visibility of the spectral colors when applied to hypsometry contributes more to cartographic confusion than to clarity. To discard this convention in favor of a more rational approach is probably impossible. It would require a generation of education.

Cartographers are not entirely to blame for their general conservatism and adherence to convention particularly with respect to lettering, color use, and map design. Private cartography, unlike some of the fields in the fine arts, has little popular support, whereas commercial cartography dominates the broad market. In highly competitive fields, and in fields catering to the well informed, improvement and change function as tangible assets. Not so in cartography. Those who buy maps or cause them to be made (school boards, teachers in various subjects, publishers, and editors), are generally quite uninformed about cartography, and as a consequence they buy what is already familiar. Since new plates, expert consultants, and new techniques are neither cheap nor do they guarantee income, the commercial map producer is naturally only too happy to peddle the older wares and reap an ever increasing rate of profit as long as possible.

The undue dependence upon convention and custom has been accompanied in many instances by absurd rationalization. Logical foundations for many of these conventions are commonly lacking. For example, Max Eckert-Greifendorff asserts that brown is the best color for terrain, contours, and land representation since "the fundamental color of the soil is brown as is especially evident in freshly tilled soil in the spring."[3] This is the kind of "logic" which has tied cartography to convention. If his "logic" be analyzed he is actually claiming that the color for landforms and isohypses should be based solely on the B horizon of middle latitude humid forest and steppe-land soils. In any case it is difficult to see why contour color should have anything to do with soil color. Contours lie on the ground and the surface of most soils is either (a) black or nearly so (and covered with green vegetation), or (b) red (and covered with green vegetation), or (c) if visible, as in arid regions, a color ranging from grays through degraded yellows. Large areas of the earth have no soil cover. How much better it would be to determine contour color on the basis of (a) maximum transparency with solid body, (b) preciseness of definition, (c) degree of continuous tone produced by lines in juxtaposition, (d) lack of disharmony with other contemplated color use, and so on. Objective investigation may point to brown as the best color, but not because it is the color of "freshly tilled soil in the spring."

Tradition and familiarity maintain a strong hold. On the other hand, it is reasonable to expect that the developments in the science of vision and the spreading appreciation of the importance of design may combine with the marked increase of interest in cartography in recent years to promote more critical examination of many of the long standing conventions.

Considering the wide use of maps and their importance as media for portraying scientific fact it is indeed surprising that there has been relatively so little written on the cartographic technique. W. M. Davis complained that "maps are . . . indispensable . . . but they are inarticulate, and their silence seems to have affected their makers. . . . It is as if their expertness in the graphic expression of facts were accompanied by an atrophy of the faculty of verbal expression following its disuse."[4] The reference to expertness may be open to question, but there can be no argument about the lack of critical written material on a large proportion of the technical aspects of cartography. Since small-scale maps are generally prepared with specific objectives in mind, it could be reasonably expected that, like any other practical creative field, such as architecture or even advertising, there would exist a body of principles and laws based on experience, experimental research, or logic which would govern the employment of the various structural materials. By reference to these principles one could, within the bounds of controversial interpretation, arrive at fairly accurate evaluations of the effectiveness with which the techniques and media accomplish their purpose.

The drawing of a parallel between cartography and architecture is instructive. Each lies in the field of the practical arts; each is older than history; and each, since its beginnings, has been more or less under the control of its consumers. The procedures of architectural and cartographic creation have been based on convention or artistic whim, and in many cases on well meant but ill-founded judgment. In general, functional inadequacies have been concealed beneath the guise of artistry, a standard form of refuge among many intellectual pursuits.

It is interesting and informative to draw such a parallel for we are witnessing what amounts to a revolution in the field of architecture. Modern building design, within the limits of conservative opposition, has become functional. It is now accepted that a structure will be planned and built according to the needs of its future users. It is not expected that the inhabitants or the weather will conform to the structure. Function provides the basis for the design. A similar revolution appears long overdue in cartography. The development of design principles based on objective visual tests, experience, and logic; the pursuit of research in the physiological and psychological effects of color; and investigations in perceptibility and readability in typography are being carried on in other fields. The more widespread use of maps, and the appearance of critical dissatisfaction lead to the conclusion that cartography cannot continue to ignore these developments, and that such a movement in cartography cannot fail to materialize.

There are indications in the literature that such a functional approach is already receiving attention. Even as far back as 1933 the National Society for the Study of Education devoted its yearbook to geography and one chapter to map standards.[5] This relatively unpretentious statement may well be the first American appeal for a truly functional approach to cartographic methodology. It specifically asks for investigations of the visual standards of maps and suggests as worthy subjects for research such topics as legibility, psychological effects of color, and clarity. Throughout the entire chapter the need for simplicity and the wider use of the special purpose map is stressed. Ten years later, during the war, a large number of cartographers in government service realized the visual inadequacies of conventional small-scale cartography, and devoted considerable thought to the subject of functionalism. A Committee on Cartography of the American Society for Professional Geographers, in drawing on that recent experience for the benefit of the academic aspects of the subject, summarized its conclusions in 1946 by pointing out that: "Greatest emphasis in a course on cartography should be placed on map design and planning as related to the purpose of the map."[6]

Although geographers and cartographers are prone to judge maps, the major portion of such functional evaluation is made, not in terms of the visual aspects, but in terms of the geographic content. It may be presumed as self evident that content is an obvious functional aspect of a map, and that neither its determination nor its evaluation is a matter of strictly cartographic method. One of the cartographer's greatest problems is the selection and generalization of the data with which he is working. How many contour lines to draw, how much to smooth them, what selection of rivers or towns to make, how to simplify a coastline, and many others, are perennial problems to the cartographer. Considerable importance has been attached to these functions in cartographic writing, and the very frequency of their appearance and concern to the cartographer (especially generalization) has fostered a tendency to think of them as cartographic. All scientific endeavor is constantly faced with the task of evaluation and generalization, and cartography, rather than being an exception, merely follows the rule.

If we then make the obvious assumption that the content of a map is appropriate to its purpose, there yet remains the equally significant evaluation of the visual methods employed to convey that content. These graphic methods, together with the logic which binds them to their function and sets the limits of their utilization, constitute the cartographic technique.

NOTES

1. O. M. Miller, "An Experimental Air Navigation Map," *Geog. Rev.* XXIII (1933), 48–49.
2. John K. Wright, "Map Makers Are Human. Comments on the Subjective in Maps," *Geog. Rev.*, XXXII (1942), 542.
3. Max Eckert-Greifendorff, *Kartographie* (Berlin, 1939), p. 41.
4. W. M. Davis, "The Progress of Geography in the United States," *Annals Assn. of Amer. Geog.*, XIV (1924), 194.
5. A. G. Eldredge, A. W. Abrams, W. Jansen, and C. M. Shyrock, "Maps and Map Standards," *The Teaching of Geography*, 32nd Yearbook of the Nat. Soc. for the Study of Education. (Bloomington, Ill., 1933), Chap. 25.
6. American Society for Professional Geographers. Committee on Cartography. "Cartography for Geographers," *The Professional Geographer*, IV (1946), 10–12.

III

Cartography as a visual technique

IS CARTOGRAPHY art? If a map has a pleasant appearance does that make it a good map? These and many related questions constantly arise in discussions of the philosophy of cartography. To champion a categorical yes or no would be presumptuous. Yet the very frequency of the questions involving the relationship of art and cartography indicates that many students and practitioners have been unable to resolve the apparent, but perhaps unreal, incompatibility between science and art.

The assumption that effective cartographic technique and its evaluation is based in part on some subjective artistic or aesthetic sense on the part of the cartographer and map reader is somewhat disconcerting. For example, E. Raisz claims that the "effective use of lines or colors requires artistic judgment,"[1] and J. K. Wright explains that the suitability of a symbol "depends on the mapmaker's taste and sense of harmony."[2] Throughout the literature there are numerous similar assertions regarding the assumed subjective aesthetic and artistic content of cartography.

There is also a considerable tendency to define the subject as a kind of meeting place of science and art. This is exemplified by Eckert. He pleads for artistic imagination and intuition in cartographic portrayal and claims that the interaction of such talents with scientific geography produces the aesthetic map.[3] There is no question about the importance of imagination and new ideas, but it is equally important that significant processes be objectively investigated, whether it be the visual consumption of a graphic technique or a process in geomorphology. In order to understand the degree to which art enters into cartography it is necessary to examine some of the fundamentals of each in order that the logic does not become simply a matter of semantics. It can perhaps best be approached by a comparison of the aims, techniques involved, and the results accomplished by each activity.

Most scientific cartography is concerned with the dissemination of spatial knowledge. The aim of visual art is more difficult to express. Generally speaking it may be said to have two basic goals, the first of which is to provide aesthetic pleasure through visual (sensuous) stimuli. In many instances, and especially significant in cartography, this has taken the relatively crude form of ornamentation. This, so far as cartography is concerned, may well be considered a low form of art based on the uncritical and popular conception that anything graphic that is difficult is thereby artistic. Even assuming (and it is a difficult assumption) that the fancy borders, ornamental cartouches, curvaceous lettering, and other decorative features so common on older maps (and still not uncommon)[4] are a source of pleasure to a reader, it does not seem illogical to suggest that such "art" does not add to the functional quality of the map. On the contrary, it may actually detract from it, for the attention may be drawn to these exhibitions of manual dexterity when it should be concerned with the data presented for consumption.[5] As with ornamentation, the use of color on maps has also been held as irrefutable proof that there is aesthetic art in cartography.[6]

With respect to the use of color, ornamentation, and other assumed elements of art, there seems to be considerable confusion between the creative motives involved and the effects produced. The fact that there is a "joy of creation" in mapmaking is certainly no argument that there is art in cartography, for there is similar pleasure associated with any creative effort whether it be painting, composition of music, or gardening. If a map is a functional object intended primarily to stimulate the intellectual aspects of our mental processes, the fact that it uses visual media to do so is beside the point. If the final composition also appears "beautiful," that aspect is quite apart from its primary function and may, as previously pointed out, detract from its effectiveness as a map, since the aesthetic response may take precedence over or interfere with the intellectual response. That is not meant to imply the extreme, that something created for practical purposes must therefore not be pleasurable. Certainly any job, well done, especially a creative undertaking, provides pleasure both to its creator and observer. What is meant, on the contrary, is that any aesthetic stimuli which may be included in a map probably should be incorporated consciously and with full realization of its effect on the other visual material.

The other aspect of art, the art which attempts to awaken various responses not necessarily of beauty, received little attention in cartography until the use of maps for propaganda purposes came into favor. It is difficult and perhaps unsatisfactory to attempt to separate the two aspects of art from the point of view of motivation because the techniques of the two are essentially similar, and certainly the motives are commonly combined or confused. In nonaesthetic art the aim may be any of a multitude of possibilities but a basic characteristic is the attempt to construct visual stimuli which will produce desired mental responses. It is well known that certain colors, shape combinations, and line relationships, produce predictable responses including intellectual connotations such as simplicity, confusion, density, rhythm, and balance.

Recently two authors have considered its manifestation in cartography and, probably as a result of reaction to its use for propaganda purposes, have taken a negative approach. John K. Wright illustrates the confusion that can occur between the visual and intellectual,[7] and Hans Speier acknowledges that "size, color, and design, can be made to serve propagandistic ends."[8] There can be little doubt that if the use of visual techniques to stimulate predictable responses is accepted as within the field of art, then cartography includes artistic techniques. Such techniques obviously should be employed in the attempt to satisfy the functional requirements of a map, for a map is a graphic thing that, by any definition, cannot be visually sterile.

The difficulty that arises in interpreting that fairly obvious conclusion is what proportion, if any, of the cartographer's artistic judgment, taste, or sense of harmony should enter into his creation. Everyone has probably heard the fatuous statement (and a great majority of us have probably echoed it) "I don't know what is good, but I know what I like." If we proceed on the basis of that grotesque admission, we find ourselves spurning maps with red on them because we don't like red; we champion conventions because they are familiar; we make brightly colored maps because they are brightly colored, whether they convey the information or not. In short, our judgment of technique is based on convention, whim, and fancy. What alternative bases can we adopt? There are two such. One is to standardize everything. Then all map readers, once they had learned the symbols and techniques would be presented only with varying combinations of familiar things. All cities always would appear as black circles and they would be named in Spartan Medium Italic type, and so on. The absurdity of such a proposition is, I hope, obvious.

The other alternative is to study and analyze the characteristics of perception as they apply to the visual presentation we call a map. They are complex to be sure, and even if the present state of knowledge concerning them is relatively scanty, logic will at least illuminate some of the broader aspects. Experience and research in other fields of visual presentation may assist. In order, however, to construct a basis for judgment it is necessary to admit that the scientific map is a functional object. Cartography is a technique, just as scientific writing or the language of mathematics, by which intellectual concepts are displayed for consumption. That from time to time a cartographer will include something for aesthetic purposes or will, by some technique or other, introduce a bit of levity on a map, does not alter the basic premise. It is only necessary that when such is done, it be done consciously and with full realization of how such procedure will affect the resulting visual complex.

Cartography for educational purposes should be considered as no more and no less creative than writing. That master of functional writing, W. M. Davis, happily extended this thought to cartography when he wrote: "It is well known that there are many geographical matters which are better presented pictorially, cartographically, or diagrammatically than verbally. Hence, it is just as important to study the proper and effective use of various forms of graphic presentation, as it is to study the values of different methods, treatments, grades, and forms of verbal presentation."[9] The ends desired thus dictate the means to be used to accomplish the purpose. The processes and functions which probably will occur between the eye and the mind of a reader must be predicted and analyzed if the technique is to be properly evaluated. To do so is difficult to say the least. In the first place, there are not yet objective data upon which to base answers to many questions involving visual-stimuli–mental-response relationships.

Part of a map is absorbed intellectually. That is to say, the visual shape and color of an object may be, through convention, completely or partially submerged in its intellectual connotation, e.g., blue water, the dot-dash political boundary, and familiar lettering. The remaining components of a map enter the brain as visual stimuli without intellectual meaning except through prior reference to a legend. Sometimes visual and intellectual qualities exist together in a single component, e.g., a large red USSR. The very adjacency of the visual forms within the map frame causes each to modify the intrinsic visual qualities of the other.

These stimuli have only recently been appreciated as having an important bearing, other than aesthetic, in the consumption of a map. True, Eckert indicated some recognition of this nearly forty years ago when he noted that "an artistic appearance, particularly a pleasing colouring, can deceive in regard to the scientific accuracy of a map."[10] But, except for color, it was not until the employment of maps as tools of propaganda that the importance of visual relationships was generally appreciated. Most of the writers on cartography have touched upon this subject, usually implicitly, rarely directly, and visual relationships are generally considered to be artistic or psychological components which cannot be avoided, but which should be guarded against. Speier echoed this sterile attitude when he wrote that "the relationships of the different lines and areas we see, the shape of parts, the distribution of colors, symmetry or its absence—all these are extraneous to the scientific purpose of a map."[11] If visual relationships are to be utilized to accomplish a positive purpose, then it becomes necessary to establish principles for their employment. The complete evaluation of cartographic methodology therefore requires that, ideally, the visual and intellectual properties of all map data, techniques, and media, be analyzed, as well as all the possible combinations of them.

From the abstract point of view all the components of a map may be placed under two basic categories, (1) cartographic data, and (2) cartographic technique. As has been pointed out previously, it must be assumed that the selection of the data to be used on a map is primarily a function of the educational purpose to which the map is to be put and should, therefore, be determined without regard to cartographic technique. This, of course, can never be true in practice since the range of techniques with which to convey geographic information is not unlimited, yet for purposes of evaluation of technique, consideration of content, wherever possible, must be removed. Its presence cannot fail to occasion responses which, consciously or unconsciously, will condition the evaluation of the technique. For this reason most preliminary advertising layouts are usually made up simply as organizations of shapes of varying value (brightness). The removal of these intellectual factors for purposes of evaluating technique is by no means as simple as it may appear at first consideration. As was noted, it is impossible in practice completely to separate data from technique, and since cartography is essentially a technique designed and existing for geographic data, it may be expected that data will play a large role in its method.

Cartographic data may be either quantitative or qualitative.[12] Quantitative data may be evaluated solely on bases of appropriateness and accuracy. The qualification of data involves problems such as its categorization and generalization, determination of ratios, and the choice of isograms and choroplethic limits. They lie closer to the substantive aspects than to the presentation. On the other hand, certain qualities of data such as comparative visibility of shapes, color relationships, and a number of others, involve problems of visual evaluation. When considered from this point of view these aspects must also be included in the evaluation of technique. Consequently, it seems logical to separate cartographic method into two general categories, (1) substantive method, and (2) visual method or technique. There is a third rubric in the complete classification of cartographic method, namely map reproduction. It acts primarily as a limiting factor on the other categories. One must always admit, however, that there is no clear and precise distinction between the categories.

A map when analyzed solely as a visual thing is essentially a group of more or less related items of all sizes, shapes, and colors: the land-water relationship, the trim page, the map proper, the legend and title boxes, relative line width, the individual words, any massing of data, color, or value areas, etc. All of these may (and should) be considered merely as items varying in size, shape and color, and bearing a direct visual relationship to one another. In the broadest classification of map components, these all may be grouped under the term "design," and would so be considered if the map were being analyzed purely as a graphic layout problem. But a map is more than that. Each of these elements of design has intellectual limitations which cannot be ignored: one is not able to organize the shapes freely; relative importance exists without regard to size and shape; and utility takes precedence over the aesthetic. A map cannot, therefore, be evaluated solely on the basis of pure design, even though considerable insight into the problems of technique can be gained by so evaluating it.

The lettering on a map is the first element requiring evaluation. Names and words are the shapes or symbols most submerged in their intellectual connotations, but they do have visual form, and as such are significant in the overall organization of the map. Their evaluation is based on two major aspects of their utility, the size and the design of the typeface. In addition, other important considerations in the methodology of map lettering include the appropriateness or suitability of the typeface, and the color of print and background and its effect on legibility.

The organization of the basic shapes within the map frame, that is to say, the structure of the map, has a significant bearing on the utility of the map. The controls which determine the possibilities of structural variation include elements of projection, balance, direction, and many others.

The third element in the visual evaluation of a map is color, used in the broad sense to include value (brightness) and intensity as well as hue. It is perhaps the most difficult of the cartographic techniques since it is a significant element in both lettering and structure as well as in its restricted consideration. Its use with other elements markedly affects their visual effectiveness. It has, however, one use more or less peculiar to it, the portrayal of categories of either related or unrelated data. Different hues, values, and intensities commonly appear in juxtaposition. It enters structural problems because its use changes the character of shapes. Thirdly, the color seems to produce significant emotional and intellectual responses.

These three visual components of cartographic technique, lettering, structure, and color, encompass most of the aspects of a map capable of evaluation from the visual point of view.

NOTES

1. Erwin Raisz, *General Cartography* (New York and London, 1938), p. 2.
2. John K. Wright, "Map Makers Are Human. Comments on the Subjective in Maps," *Geog. Rev.*, XXXII (1942), 542.
3. Max Eckert, *Die Kartenwissenschaft*, 2 Vols. (Berlin and Leipzig, 1921, 1925), Vol. II, Part 4, especially pp. 670–677. See also F. Becker, "Die Kunst in der Kartographie," *Geographische Zeitschrift*, XVI (1910), 473–490.
4. See The World Map, Washington, 1943. The Cartographic Dept. of the National Geographic Society.
5. Cf. Max Eckert-Greifendorff, *Kartographie* (Berlin, 1939), p. 27.
6. Ibid., pp. 31 ff. Yet elsewhere Eckert has made a searching analysis of the color research carried on around the turn of the century by Peucker, Bruckner, et al., and approaches it from a strictly objective point of view.
7. Wright "Map Makers Are Human."
8. Hans Speier, "Magic Geography," *Social Research*, VIII (1941), 314.
9. W. M. Davis, "The Colorado Front Range," *Annals Assn. of Amer. Geog.*, I (1911), 33.
10. Max Eckert, "On the Nature of Maps and Map Logic," trans. by W. [L. G.] Joerg, *Bull. of the Amer. Geog. Soc.*, XL (1908), 347.
11. Speier, "Magic Geography," p. 313.
12. The segregation of cartographic data into either quantitative or qualitative categories is generally recognized. For example see Alfred Hettner: "Die Eigenschaften und Methoden der kartographischen Darstellung," *Geographische Zeitschrift* XVI (1910), 75–82, and Wright, "Map Makers Are Human." A very clear and concise comparative treatment of quantitative and qualitative values as applied to symbols (with chart) appears in John K. Wright, "Cartographic Considerations," *Geog. Rev.*, XXXIV (1944), 649–652, Appendix I, A Proposed Atlas of Diseases.

IV

The importance of lettering

ALMOST all maps contain lettering. Yet few cartographers are entirely pleased with the lettering, and together with color it is frequently the butt of the criticism leveled at any map. The lettering is said to be too big, too small, ugly, poorly executed, too heavy, too light, poorly printed and so on *ad nauseam*. This almost negative attitude is curious, for we do not find a similar attitude toward many of the other techniques. Projections are a constant problem to the cartographer, but we do not find, for example, anyone suggesting that the ideal map is one without a projection. One of the foremost cartographers in the United States presents this concept unequivocally when he states:

> The application of lettering is one of the most difficult problems in cartography. The essential of the problem derives from the fact that lettering is not a part of the map according to our definition, for it is not visible on a conventionalized picture of the Earth; but it is a necessary addition to identify features.

The names by their bulk cover up many of the important elements of the real landscape and prevent the reader from seeing the map as a picture of the earth. On small-scale maps city names often cover hundreds of miles in length, even if printed in the smallest readable type, and their least disturbing placement is a trial to cartographers. The development of expressive cartography has been hindered more by lettering than by any other cause. [1]

This attitude toward lettering, although undoubtedly extreme, is a reflection of the modern trend toward self-expression. [2] There would seem to be reasonably sound theoretical bases for this view with respect to topographic maps of very large scale, but even in such cases the gross assumption that all places and symbols can be made self explanatory or will be known to the reader is undoubtedly erroneous. To put it simply, cartography is a medium of presentation for spatial data and it follows that when such data requires identification, then that identification becomes an integral part of the map. [3]

The identification of data and locations has always assumed an important place in cartographic technique. For a great many maps, authorship and period may be determined merely by glancing at the lettering. Maps of the Royal Geographical Society, of the nineteenth and early twentieth centuries, and many others, are all clearly identifiable by their lettering, and the frequency with which it is possible to recognize the source of lettering is an indication that it takes a prominent position in the scale of visual characteristics. Although no tests have been made, so far as we know, it is reasonable to postulate that, for most small-scale maps, the first reaction of the reader, consciously or unconsciously, is to the lettering. In many cases, of course, the lettering is even necessary for the identification of the area mapped. In addition, lettering with its background usually presents the greatest value contrast on the map, and value contrast is one of the more important elements of perception. Although the sharp angles and complex curves of letter forms make them one of the most complicated visual elements, much of this complexity escapes unnoticed since the shapes of letters are well known to the reader. Even though familiarity reduces the effect of visual complexity, the inherent shape differences between the lettering and the other line work on the average map is yet another reason that the lettering assumes a relatively important place in the visual scale. Only on a few maps, such as physiographic, or land-type diagrams, or maps of intricate coastal areas, does the complexity of the line work rival the lettering.

The evolution of cartographic lettering is an interesting study of the interaction of art, tradition, and convention. For the past several centuries an additional influence has been the method of reproduction. So long as copper engraving was the principal process of reproduction there was no problem in reproducing the finest of lines and a progressive degeneration of typestyles took place on maps, as for example, the loss of all semblance of proportion between the thin and thick strokes in Roman. Type design improved rapidly after the general revolt against the mid-Victorian typestyles, but cartographers were slow in taking back the initiative of styling the lettering for maps, being content to allow the engraver to determine the style. The practical problems of printing hairline work from the grained surface of zinc plates, together with the preparation of copy for photography rather than engraving, caused the general problem of map lettering to undergo review during the first half of the present century. Unfortunately there has been no extended research into the problem of styling lettering for maps, and the present employment of faces shows neither regard for some of the more important typographic principles nor functional consistency. The function of lettering on small-scale maps differs from that on large-scale and reference maps. In the latter case the names are employed as a reference to be used when desired. A good reference map will naturally employ a great many names and, in general, the criteria of selection and use of type will be modified by the desire to subdue them in order that "their bulk (will not) cover up many of the important elements."[4] However, this does not apply with respect to small-scale special purpose maps wherein the entire map is prepared with a limited purpose in mind. Here the requirements are the construction of a clear, homogeneous, and legible presentation of some specific data. Withycombe in 1929 summed up the essentials of typography for maps as follows:

> The essentials to aim at are . . . 1. *Legibility.* The letters not only must be legible when standing alone but also when superimposed upon the detail of the map. 2. *Suitability for reproduction by the photographic process which is to be used* . . . 3. *Good style and intrinsic decorative qualities.* The style of the lettering on a map should be as good as that exhibited by the best fonts of type in use by book printers. As legibility is one of the characteristics of every really good alphabet, the first aim will be attained if really good style is achieved. 4. *Distinction and contrast.* Certain classes of names should be clearly distinguished and the different types of alphabets used and their gauge [point size] and spacing should achieve this. 5. *Harmony of effect.* The alphabets appearing on any one map should harmonize with each other, [and] with the detail of the map. . . . [5]

His analysis includes most of the factors to be considered in evaluating the lettering on maps, as well as one which is no longer quite so important as it was when he was considering the problem. In the decade following the publication of Withycombe's paper, many strides were made in reproduction techniques, and during World War II even more progress took place. A noted typographer and expert in the field of typography for lithography has stated categorically, "any typeface can be reproduced beautifully by photo lithography if the workmen . . . know their trade and do the best they can."[6] Nevertheless, reproduction is rarely perfect and the cartographer can hardly ignore the possibility of the lines breaking or openings filling with ink and the other hazards of reproduction by which the lettering of a map must pass.

Before examining the bases for evaluating lettering techniques it would be well to point out that a great many maps, particularly special purpose maps, during the early part of this century were hand lettered by the cartographers, draftsmen, and engravers. As few of these craftsmen can be termed experts in type design it is to be expected that a great many maps contain typefaces which defy classification. The manual entitled *Topographic Instruction of the United States Geological Survey* in directing the style of engraving lettering specifies the faces only by such terms as "block," "stump," "roman," and "italic capitals."[7] The Ordinance Survey of the United Kingdom uses similar language.[8] An American text employs such terms as "inclined gothic" and "italics."[9] Some federal agencies, among them the Army Map Service, the Bureau of Plant Industry, and the State Department have used commercial typefaces for stickup. So far as it is known, only the National Geographic Society uses specially designed typefaces.[10] The use of commercial type and the stickup process is rapidly gaining favor. Prior to the past fifteen years the stickup process was laborious and the results not easily reproduced;[11] but recently, many techniques for application have been developed, and it is to be expected that in the future stickup will become the accepted process for lettering on most maps. The problem of selection of typefaces will thereby be enormously increased because of the thousands of typefaces from which to choose.

The technique of lettering on maps covers a wide range. Perhaps the first question of choice facing the cartographer is that of the form of the typeface. There are an infinite number of possibilities from freehand lettering to the innumerable styles available from printers and typographers. They vary in legibility, appropriateness, texture, and even in the general character or mood they represent. Next the cartographer must decide on sizes for, after all, the best typeface is of little concern if it cannot be read. The relative sizes are of great significance in a map in terms of comparative emphasis and legibility. Inherent in the above questions is that of the color of the lettering and of the background on which it appears, for this constitutes one of the major controls of legibility.

These, and some other aspects of the lettering technique have been investigated by a number of researchers, but unfortunately they have not been cartographers, without significant exception. Consequently their determinations are not precisely applicable to cartography but their work has resulted in some general principles which can be used as a guide to lettering technique in cartography.

NOTES

1. Erwin Raisz, *General Cartography* (New York and London, 1938), p. 156.
2. Cf. Eckert-Greifendorff, *Kartographie* (Berlin, 1939), p. 26.
3. Cf. Eckert's reaction to the *Stumme* map (*Die Kartenwissenschaft*, Vol. I, 347), when he quotes from Peucker, "the lettering belongs to the map, as speech to men."
4. Raisz, *Cartography*, p. 156.
5. J. G. Withycombe, "Lettering on Maps," *Geog. Jour.*, LXXIII (1929), 429–446, ref. p. 432.
6. M. T. Monsen, "Importance of Type in Lithography," *The Lithographer's Manual* (New York, 1940), p. 72.
7. C. H. Birdseye, "Topographic Instructions," *Geological Survey, Bull. 788*. (Washington, Dept. of the Interior, 1928), pp. 332 ff.
8. Withycombe, "Lettering on Maps."
9. Raisz, *Cartography*, pp. 156 ff.
10. The National Geographic Society has designed its own typefaces. A photographic device is used which allows them to "compose" photographic negatives as type in a frame for exposure. The resulting positive is made adhesive and is used as stickup. The result gives the appearance of freehand lettering with the precision of type. See Wellman Chamberlin, *The Round Earth on Flat Paper* (Washington, 1947).
11. Cf. American Society for Professional Geographers. Committee on Cartography, "Cartography for Geographers," *The Professional Geographer*, IV (1946), 10–12; and Eckert, Die Kartenwissenschaft, II, 681.

V

The style of lettering

CARTOGRAPHIC calligraphy is rapidly becoming a thing of the past, for the freehand is being replaced by the type letter. The use of metal and other sorts of type raises a number of questions for the cartographer, not the least of which is the form or style of the typeface. Quite apart from the aesthetic qualities of type designs, it is to be expected that various styles will be more or less appropriate, legible, perceptible, and so on through the list of cartographic functions and qualities applicable to lettering.

Since the styles of lettering and type have evolved over a long period of time, it is only natural that there are many different designs and that the various different kinds of lettering are as hard to classify as climate, for the elements of design shade gradually from one style to another. It is possible, however (with a little squeezing), to group the major kinds of faces into categories.[1] The oldest of the designs is appropriately called Old Style or Classic Old Style. It originated around 1500 from Aldus and evolved primarily through the work of Garamond. It is characterized (Figure 1) by subtle differences between the thick and thin lines and by its lack of geometric precision. The serifs are usually flowing and slanted. In great contrast to the Old Style is the so-called Modern (Figure 1) usually associated with Bodoni around 1800 but probably much earlier. The Modern type is quite different from Old Style in that it looks geometric and precise as if drawn with a straight-edge and compass. The serifs are hard, straight and of uniform thickness. There is great contrast between the thick and thin lines of the Modern, and some of the Moderns of the nineteenth century overdid this to a great degree. The Sans Serif style (Figure 1) is a twentieth century innovation first developed in Europe. It has the geometric quality of the Modern but has, of course, no serifs and there is little if any difference in thickness of strokes.

This line is set in 14 point Caslon

This line is set in 14 point Bodoni Book

This line is set in 14 point Metrolite

This line is set in 18 pt. Kaufmann Script

This line is set in 14 point Cairo Bold

This line is set in 14 pt. Engravers' Old English

Figure 1. Examples of classification of typestyles. From top to bottom the groups are: Classic Old Style, Modern, Sans Serif, Script, Square Serif, and Text. (Cf. De Lopatecki.)

The other three kinds of typefaces are less commonly seen on maps (which is fortunate), except that there has been a tendency in the past to employ the Square Serif. Text and Script are rare in type but occasionally are seen in the freehand form. The use of type in cartography is similar to its use in advertising where the subject of type in display has received considerable attention. The purpose of type in advertising display is not exactly similar since considerable emphasis is placed on the aesthetic aspects of the various faces, but good typefaces artistically are usually good functionally. Aesthetics enter into the question when one contemplates employing the fancier faces. As Reeves forcibly stated in the discussion following Withycombe's paper, "we do not want to attract attention to the names, but to the map features."

The classification of typestyles is capable of no more precision than the classification of many other phenomena in geography and cartography. Sometimes it is difficult to decide in which rubric to place a hybrid. In addition, in recent years various devices have been developed to aid the unfortunate who has not the elemental manual dexterity to letter freehand or who refuses to learn how to do so. For the few and far between maps which come from the table of this sort of individual, stickup is not appropriate because of its expense. Consequently, he leans on the mechanical crutch.[2]

The chief function of type is to be read, and legibility of type is a subject with many complexities, of which the style of the typeface seems to be a less important consideration compared to some other aspects treated later. However, legibility definitely does vary with typestyles, and a number of studies have been made on this subject. These may be divided into two categories, those concerning legibility, and those concerning perceptibility; both show results of interest to the cartographer.

Comparative legibility of typestyles is of importance in cartography in the selection of faces to be used in the following ways: (a) blocks of type such as legends and explanatory text; (b) place names and other material set closely, as it would be in ordinary reading; and (c) words, necessary on the map but subordinate to the main theme. Comparative perceptibility is of importance in the selection of faces to be used in the following ways: (a) names widely spaced, (b) unfamiliar names, and (c) names and words requiring different emphasis. In tests conducted by Paterson and Tinker various typefaces were employed in identical reading tests and the results arranged according to reading time.[3] The faces selected are, by chance, reasonably representative of some of the classifications here made. If the results are tabulated according to these groups of type, the Classic Old Style leads in legibility closely followed by the Modern and Sans Serif. The amount of difference among these groups is small enough to be immaterial. It is interesting to note that a typewriter face shows a significant departure by being upwards of 4 percent slower than Classic Old Style, while the Text type retarded reading by over 13 percent. In a further test reader opinions were gathered and ranked according to the legibility of the same typefaces. As might have been expected there was little agreement between the two tests except that the Text type fell at the bottom of the list in each case. It is also interesting to note that the Sans Serif, although showing no marked effect on legibility, was thought to be distinctly less legible.

Generalizing from the above data, meager though it is, it is possible to observe some significant results for cartography. Perhaps most important is the inference that apparently there is no marked difference between any of the standard types in the Classic Old Style, Modern, or Sans Serif groups. Consequently, it is safe to say that any typeface in these groups which has stood the test of time will be equally legible for cartographic use. Choice of a face in the Sans Serif group should, however, be accompanied by a realization that reader opinion considers it less legible. It is doubtful if anyone would actually say "this typeface is less readable," but the reason for the reader opinion as expressed in the test would be interesting. No explanation was offered by the testers. The Sans Serif group is a relatively modern typeface. Its first representative, Futura, was introduced in this country within the present century and this face, together with the others in the group, show marked departures from the Classic and Modern group. Although somewhat related to a few of the Square Serif faces it appears quite different. Sans Serif is a radical departure from the commonly used faces and as such is bound to attract attention, a quality which may or may not be desirable in a particular cartographic composition.

In one of the first (and most complete) studies of perceptibility Roethlein showed that there are definite differences among various styles of typefaces, as well as among the various letters in each face.[4] Of more significance was the conclusion that perceptibility depended to a great extent, on the thickness of the line. It was found that perceptibility increases with increasing thickness up to a point and then decreases. The studies were not extensive enough to reveal what the optimum is.

During the thirties Luckiesh and Moss developed a Visibility Meter and tested a number of typefaces.[5] From the visibility (perceptibility) ratings for each face they calculated the point size ratings for equal visibility. The resulting table (Table I) is of considerable interest since it seems to bear out Roethlein's contention

TABLE I

TYPEFACE	PERCENT VISIBILITY (Bodoni Book 100 Percent)	POINT SIZE FOR EQUAL VISIBILITY
(8 Point)		
Bodoni Book	100.0	8.0
Bodoni Italic	96.0	8.3
Bodoni Bold	108.3	7.4
Caslon Light	96.2	8.3
Caslon Light Italic	81.1	9.4
Caslon Bold	106.5	7.5
Sans Serif Light	97.1	8.2
Sans Serif Medium	105.7	7.6
Sans Serif Bold	104.3	7.7
Cheltenham Wide	100.5	8.0
Cheltenham Bold	108.5	7.4
Cheltenham Bold Condensed	93.2	8.5
Light Copperplate Gothic	98.1	8.1
Heavy Copperplate Gothic	102.5	7.8
Goudy Light	94.0	8.4
Goudy Antique	106.2	7.5
Goudy Bold	104.8	7.7
Cochin Light	102.3	7.8
Cochin Bold	112.6	7.1
Garamond Bold	118.7	6.6

that perceptibility is in large part dependent upon thickness of line. Bodoni Book type was used as a standard, and an analysis of the table shows that all bold variations are (1) over 100 percent visibility and (2) an average of over 8 percent more visible than the standard. It also shows that the most variation occurred in the Old Style and Modern faces and the least variation in the Sans Serif. Their calculations show a variation in visibility between standard and bold faces of approximately 1 point size.

There seems to be ample evidence that typefaces do vary in their effectiveness and that unless determinations of readability and perceptibility are considered in the selection of faces for maps, considerable variations in visual effectiveness may result.

Another practice, common in cartography, is the use of italic or slanting type. Italic faces were employed frequently during the early days of printing, but slanting letters have never enjoyed much popularity in the printing trade, and since as early as 1550 italic letters have been reserved for special uses. In cartography the choice between upright and italic is usually based solely on the desire to distinguish categories of data by a change in typeface. Although it is questionable whether this convention can be justified, it is interesting to note that no work on cartography questions the use of italic or slanting lettering, while most works on advertising categorically state that italic type is generally undesirable except for emphasis.

Here again, as is the case with many cartographic techniques, no real study of the comparative legibility of italic versus upright type as used on maps has been made. A near approach is a study in legibility by Paterson and Tinker who determined the reading speed of several hundred persons with material set in upright and italic.[6] The results are surprising since there was little difference in reading speed. However another test showed that reader opinion was definite that the italic was more difficult. Although reading speeds for the two styles were not far apart Paterson and Tinker emphasize that there is a definite "possibility that italics may involve greater discomfort and greater expenditure of effort." Accordingly they make the following recommendation: "The use of italics should be restricted to those rare occasions when added emphasis is desired." The study by Luckiesh and Moss, previously referred to, also throws interesting light on the subject. Their study of visibility (perceptibility) as contrasted with legibility shows (Table I) that the italic faces consistently ranked less than 100 percent visibility. From an analysis of the table it may be seen that the only faces which ranked below 100 percent in addition to the italics, were light faces. This leads one to believe that it may be the traditional lightness of face in italic that retards the visibility rather than the design. At any rate, this must remain conjecture since no comparable tests are known. Certainly, the conventional use of italic or slant letters for water features handicaps these features with respect to perceptibility.

Perhaps the next question most commonly met in cartographic lettering, after the face has been chosen, is whether to use capitals or lower case for names. Capitals are usually employed for the larger names, such as countries, oceans, and continents, and for such items of importance as principal towns, capitals of countries, and mountain ranges. Smaller towns and features of lesser magnitude commonly appear in lower case. This practice probably stems from the reasoning that capitals are more prominent, as well as from the unfamiliarity, in ordinary reading practice, of seeing large type set in lower case. Reasons for these conventions are not given in the literature, and an analysis of letter forms makes it doubtful that the convention is particularly well founded. Since words in capital letters take up more space and are therefore more difficult to focus upon than words in lower case, it is to be expected that lower case combinations would be easier to recognize.

Paterson and Tinker note: "The decreasing use of 'all capitals' in newspaper headlines and in advertising copy suggests that printers and advertising experts are coming to believe that lower case text is more legible."[7] Finding no satisfactory evidence upon which to base a determination of this question, they performed two tests on a total of over five hundred persons. The tests showed that lower case was read almost 12 percent faster than material set in all caps, and that reader opinion was *nine* to *one* in favor of lower case being more legible. In view of this evidence they recommend that "the headline writer and the advertising copy writer should henceforth abandon all capitals and should rely primarily on lower case." On the other hand the difference in legibility may not be so great as to make it undesirable to use capitals for certain categories as, for example, different population categories or political divisions of differing magnitudes. It would seem that different combinations of point size, capitals, capitals and lower case, and capitals and small capitals would be preferable to a change in typestyle.

Although there is no question that words in lower case are more legible than all capitals, it should be noted that much of the lettering in cartography is letter spaced to the extent that it is stretching a point to consider the arrangement a word. Actually, such letter spaced words consist, visually at least, of isolated letters. There is every likelihood that there is considerable difference between the perceptibility of isolated lower case and capital letters. However, no known tests have been made. Roethlein's tests are not applicable since size of the letters was not controlled so that they might be comparable.

In addition to the primary function of legibility, type on a map by its visual forms (and size) creates a number of impressions regarding texture, contrast, and mood. An examination of a large number of published maps ranging from soils maps to atlas maps shows the use of a maximum of twelve different typefaces and a minimum of eight. It is believed that this is representative of most special purpose maps, and it illustrates one of the primary cartographic conventions, the practice of distinguishing the different features of a map with different styles of lettering. The epitome of this tradition seems to have been reached in the international map, the style sheet of which shows at least twelve different faces of type. Hinks, in writing of this tendency, deplores the emphasis placed on variety by noting: "A great part of this deliberate variety is superfluous, since there can rarely be any doubt about the character of the feature to which the name is attached, if the name is suitably placed."[8]

Although he apparently was unaware of it, Hinks' general opinion is concurred in by all writers on type display and typography.

De Lopatecki categorically states: "As a guiding principle, use only one kind of type with its variations," and again, "to attain typographic harmony . . . use only variations of one face . . . and . . . use as few of these, both in matter of weight and size, as possible."[9] There seems thus to be a fundamental disagreement between the typographic experts and the cartographers, and one cannot help but lean toward the view of the typographers since they have had considerably more experience than cartographers in the field of type use and design. All designers, regardless of the media with which they work, accept as one of the first principles of visual structure that all elements of a visual composition should be in harmony. This does not, by any means, obviate the use of contrast, but it does mean that contrast should be gained by variation of one or more of the elements of contrast without absolute opposition such as would be obtained by using diametrically opposite typefaces such as one face from the Old Style group with a member of the Modern group.[10] The assertion that a different face should be used for each feature or group of features is dubious from a visual point of view, and it would appear logical to accept the principles of typographic harmony unless a justification for a departure from these principles can be made.

The principles of typographic harmony are primarily based on the shape of the letters and have developed through years of use. They are stated only generally, probably because of the tremendous variety of typefaces, but the basic principle and the manner in which it is usually found is well illustrated by the following statement from Frazier: "Everything considered, the results which are the most satisfactory are usually found in the printing in which the question of the association of typefaces does not enter, the printing in which but one series of type is used."[11] This does not work much of a hardship since most common faces have at least six variants, such as light, standard, bold, extra-bold, italic, and bold italic. De Lopatecki lists a few possible combinations in addition to the general principle: (1) A strong geometric Modern (Cf. Bodoni) with a Sans Serif; (2) Square Serif with Sans Serif; (3) Sans Serif with Old Style.

The inescapable conclusion is, that until much more objective studies of typographic harmony are made, the principles of the typographic experts must be followed even if they are primarily subjective. They have, however, stood the test of more frequent and wider usage than has been the case with cartographic lettering.

Under the heading of *Harmony of Effect* Withycombe stated that the "alphabets appearing on any map should harmonize with the detail of the map." By this he referred to two other aspects of the fitness of lettering, its texture, and its mood, or in other words, its appropriateness to the subject matter or purpose of the map. Again, the bases for lettering methodology in these connections are derived, as in typographic harmony, primarily from subjective experience, and typographic practice. A typographer concludes: "In selecting a type to be used with line drawings it is necessary to recognize some definite quality in the style or execution of the drawing that may act as a guide in selecting the appropriate design of typeface. [12]

Texture of type is a relatively clear concept. Different fonts have characteristic shape features which may either harmonize well or contrast well with the characteristic shapes on a map. Contrast of shape is one of the more important elements of visibility, and if a map is constructed functionally, the degree to which the lettering should stand out or be subdued may be markedly affected by choice of typeface. A light face Script type used on a line physiographic drawing will blend into the lines on the map, because of the similarity of shapes. This may or may not be desirable. A Modern face, such as Bodoni in which the contrast between thick and thin strokes is great, will appear lost on a map with a pattern of lines basically similar. A map with a strong geometric quality would require a typeface of rounded design such as Caslon or Goudy if the lettering were to stand out, or Sans Serif if it were to be subdued.

Of all the aspects of fitness of type on a map the methodology of mood is the most subjective. Yet, the existence of such an element in typefaces cannot be doubted. Poffenberger and Franken made a series of tests involving the association of common objects and connotations in advertising with typefaces and conclude that "the results of the experiment show quite conclusively that different typefaces do vary in appropriateness."[13] De Lopatecki points out that "typefaces all have individual personalities" and joins to various typefaces such adjectives as: austere, elegant, precise, modern, and dignified.[14] Although the element of mood should unquestionably be included in lettering methodology there is nothing in the literature to give guidance. In any case, as Poffenberger and Franken point out, the effects produced by typefaces must be extremely mild. Further study is clearly required before definite principles can be evolved.

One of the most important problems of the cartographer relative to typestyles is that concerning the effects of reduction. Types are separately designed for each size within a font, and reducing a size changes the size/line-width relationship (usually making the face lighter). So far as is known by this author, no movable types have been designed for cartographic use or for any use involving reduction.

NOTES

1. See Eugene De Lopatecki, *Advertising Layout and Typography* (New York, 1935). Classification of typefaces is not easy because of the subtlety of the differences in the gradation from one group to another. Generally speaking, writers on the subject have avoided classification based on visual characteristics alone and have grouped the faces variously on the basis of mechanical composition, designer, and period.
2. Most of the lettering produced by these devices would be classified as poor examples of the Sans Serif, although there is a new product on the market which has possibilities for improving the quality of this kind of mechanical writing.
3. Donald G. Paterson and Miles A. Tinker, *How to Make Type Readable* (New York, 1940), p. 16.
4. B. E. Roethlein, "The Relative Legibility of Different Faces of Printing Types," *Publ. of the Clark Univ. Lib.*, III (1912) 1–41.
5. Matthew Luckiesh and Frank K. Moss, "The Visibility of Various Typefaces," *Jour. of the Franklin Inst.*, CCXXIII (1937), 76–82.
6. Paterson and Tinker, pp. 20–22.
7. Ibid., pp. 22–23.

8. Arthur R. Hinks, *Maps and Survey*, 2nd. Ed. (Cambridge, Cambridge Univ. Press, 1942), pp. 46–47.

9. De Lopatecki, *Advertising Layout*, pp. 78–80.

10. Ibid., p. 81.

11. J. L. Frazier, *Modern Type Display* (Chicago, 1920), p. 42.

12. Cf. Herbert Jones, *Type in Action* (London, 1938), p. 75.

13. A. T. Poffenberger and R. B. Franken, "A Study of the Appropriateness of Type-faces," *Jour. of Appl. Psych.*, VII (1923), 312–319, ref. on p. 328.

14. De Lopatecki, p. 83.

VI

The employment of lettering

THE application of the chosen typestyle or styles to the map requires that a number of decisions be made, beyond that of style. These questions are similar to those involved in the selection of the typefaces, for they also are primarily functional in nature. To return to the analogy of architecture, the lettering on a map is like some critical building material. Before the kind of material can be chosen it must be evaluated in terms of its properties as to appropriateness and ability to do the job under consideration. After the material has been selected, the best way to utilize it remains to be determined. Similarly we find the cartographer who has chosen Goudy Old Style and Futura Medium being then faced with such questions as the proper size to employ, the effect of color of print or background, and the orientation of the lettering.

Perhaps the most common decision which must be made by the cartographer concerns the size of lettering to be used for the great variety of items which must be named on maps. Specifications for lettering are usually based on the size of the item or the space to be filled. It is also considered standard procedure to increase the size of the lettering according to the relative importance of places or things. Since a great many maps contain a large number of names, it may be expected that in general their size would be small. It is important, however, that the number of names be not so great that the smallness of type becomes unreadable. Like any material designed for visual consumption, "insofar as any data on a map cannot be grasped by the eye, and easily read, in just that measure is the map encumbered with useless material, and is a failure."[1]

Type size is always designated by points. One point equals approximately one seventy-second part of an inch. The points refer to the body of the type and not to the letter height and it is obvious that the letter will always be somewhat smaller than the point size. There is also, of course, some difference of size among the many styles of type.

In order to gain some idea of the standard practice in atlas production, twelve well-known atlases were examined and the maximum and minimum lettering sizes estimated and tabulated. The average size of the smallest lettering is 3–4 point and the average of the largest is 10–12 point.[2] It is evident that there is not much range and that the smaller sizes are small indeed. Most special purpose maps conform to atlases in the size of their lettering although in general the smaller number of names on this kind of map makes it possible to use larger sizes for the smaller names.

Cartographers often speak of legibility in connection with lettering and complain about the smallness of lettering on maps. At present the reading glass is a necessity for those who frequently consult atlases.

As in other aspects of the methodology of lettering, type size varies both in legibility and perceptibility. In a series of experiments Paterson and Tinker found speed of reading to be markedly affected by size of type. They point out that, for reading, "type sizes smaller than 8 pt. or larger than 12 pt. are quite unsatisfactory."[3] Although the reading of a map and the reading of a printed page are two quite different operations, it may be expected that there is at least a degree of similarity between the two. If excessive fatigue results from reading type smaller than 8 point in a book, it follows that similar fatigue will result when one attempts to find a particular name or set of names on an atlas page if the majority of the names are set in type smaller than 8 point. The upper limit of easy legibility is of little concern since the large names on a map are not read as type in a book.

Of considerably more importance to cartographic methodology is the study made by Luckiesh and Moss on the quantitative relationship between visibility (perceptibility) and type size.[4] In their experiments a kind of common denominator of perceptibility was established by reducing the amount of contrast between type and the background on which it was printed to a point at which the type was invisible, and then increasing the contrast until the type could be recognized. In general it may be considered as a score of relative visibility based on the angle subtended at the eye by objects (which is the basis for the normal testing of eyesight), or conversely, relative visibility. Using Bodoni Book typeface the following scale of relative visibility was determined.

TABLE II[5]

SIZE IN POINTS	RELATIVE VISIBILITY
3	1.10
4	1.60
5	2.11
6	2.64
8	3.64
10	4.65
12	5.66
14	6.67
18	8.67
24	11.68

Even a casual comparison of the atlas type sizes and the figures presented in Table II and Figure 2 clearly indicates that current cartographic practice needs some revision with regard to type size. Luckiesh and Moss, using accepted medical techniques, conclude that the threshold of normal vision is between four and five point type, yet standard cartographic procedure calls for a similar size for the majority of lettering on maps. Clearly the lettering sizes on maps have not been selected on the basis of readability. In addition to the problem of legibility, techniques have now been devised which give the cartographer an opportunity to select sizes of type in relation to the relative emphasis he desires to assign to names.

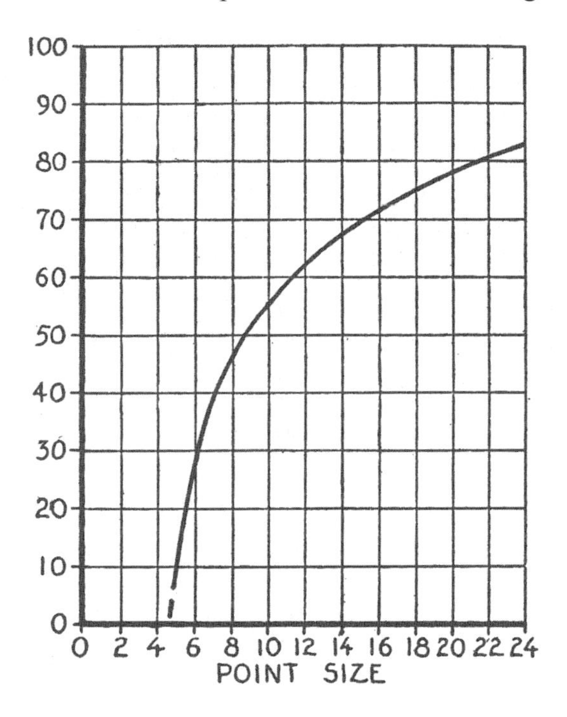

Figure 2. Percentage maximum visibility according to point size. (From Luckiesh and Moss, p. 94.)

Luckiesh and Moss have in their studies presented some of the fundamentals of the lettering technique as it applies to cartography. Their studies, of course, concern only one style of type. Similar curves could be worked out for all others likely to be used, as well as tables of proportion one to the others. At any rate, bases for the functional use of type sizes have been established and only future research is necessary to provide the data for the effective evaluation of the methodology with respect to type size. The reader is referred to an example by the author of the practical results obtainable from this kind of research.[6]

The orientation of lettering on maps seems to be, in large part, based on the convention of placing north at the top of the page. Although the basis for north orientation is obscure it is quite understandable why the corollary orientation of lettering should be considered desirable. If lettering were placed without reference to visible grid lines a confusing pattern would be set up between the regular grid and the irregular eye movements caused by reading the names at competing angles. With the exception of some survey and boundary lines the projection grid as shown on most maps is by far the most regular shape. It therefore sets the pattern for the placing of all other movable features including the lettering. So long as the projection is one of the rectangular or conic variety, it is simple to orient the lettering with the parallels, if the map is designed with north at the top. The azimuthals present another problem!

The fact that most type has been designed to be used in horizontal form is probably the most important element of choice in the orientation of lettering. Letters are very complicated visual forms and their designing is an exacting process. Each letter must "fit forward and backward" with each other letter both in capitals and in lower case. The shapes are adjusted many times before the letter is cut and it is therefore not surprising that when the letters are presented on a nonhorizontal or curved base they attract attention by their unharmonious position and appearance. This is particularly noticeable along rivers where it is necessary for the lettering to follow the sinuosities of the stream in order that the eye movement be similar. Flat based lettering would be out of place. Very few letter forms have been designed especially for cartographic use, and it is doubtful if those that have were devised with this problem in mind.

Many cases occur on maps where crowding requires that lettering be curved. It has been shown by Paterson and Tinker that the upper half of both capitals and lower case contain more clues to letter and word form than the lower half.[7] It would seem logical that when curved lettering is necessary, separation of the lower portion of the word, rather than the upper, would aid in legibility.

The ideal criteria for the orientation of map lettering, like those for other components, are based primarily on utility. This includes degree of emphasis desired and eye movement. Since most lettering is designed to be read horizontally it follows that anything not horizontal will attract attention by being somewhat confusing. This may be either positive or negative to the purpose of the map. If the orientation of the lettering is carefully standardized in a regular pattern the map appears stable and authoritative, regardless of its substantive quality. If the lettering is oriented variously the opposite effect is produced which may seriously impair reader reaction.

One of the lettering problems commonly faced by the cartographer is the color of the print and the background. The fact that there will be differences in legibility and perceptibility between various color combinations is evident; and the problem is not a minor one since it is hard to imagine color on a map without lettering thereon. Surprising as it may seem there is almost no mention in cartographic literature of this significant aspect of lettering methodology.[8] With the increase in printing techniques and the associated construction techniques it is not difficult today to employ color and if the cartographer is interested in the visual effectiveness of his map (as he should be), color of print and background is at least as significant in cartography as it is in advertising. Since advertising is more commercial in nature, has attained importance more recently, and is less governed by convention, it is natural that the first experiments should have been reported in books on that subject.[9] Since the beginning of the century a number of experiments have been conducted and published, and the results clearly indicate that color of print and background is a subject of considerable importance to cartography.

The procedural aspects are complicated. In addition to the variations of color of print and background, the possibilities of which are many indeed, factors such as size of type, style of typeface, and others also influence the choice. Unfortunately the tests which have been made are meager, although they show results of significance. They are not, in all cases, directly related to the problems of cartographic technique since legibility and perceptibility of maps are not the same as for ordinary reading practice.

The development of the special purpose map and mechanical screens during the past few decades, together with the employment of film negatives rather than glass, has resulted in the widespread use of white or open lettering on black, gray, and colored backgrounds.[10] The use of this technique has been most common on the black and white map, both because the majority of special purpose maps are printed without color and because of the simpler reproduction techniques. What appears to be the first study of the perceptibility of black on white versus white on black showed black on white to be more easily seen.[11] Later studies and interpretation along the same lines, as cited and interpreted by Paterson and Tinker gave controversial results.[12] In an attempt to arrive experimentally at an answer to this question Holmes and Taylor each conducted perceptibility tests. Holmes found that black on white could be seen at a distance almost 15 percent farther than white on black.[13] Taylor, employing a variety of tests, found that in every case black on white is more readily perceived than white on black.[14] It is interesting to note that Taylor found Kabel Light in the larger sizes, a Sans Serif typeface, to be perceived as readily in either fashion. In a study of legibility a 42 percent advantage in reading was found for black on white over white on dark gray.[15] Paterson and Tinker measured the speed of reading with a larger number of persons, and they report a 10.5 percent advantage for black on white. Reader opinion as reported by them was overwhelmingly in favor of black on white.

The above evidence indicates without serious question that black on white is more easily read and perceived than the converse. However, as applied to cartography, it indicates only that white lettering is less efficient than black lettering *of the same size*. If it is necessary or desirable on a map to use reverse or open lettering together with black lettering of the same size, the open should accordingly be utilized for data of secondary importance or some other adjustment made to equalize their effectiveness.

A number of studies of legibility and perceptibility of black and colored print on colored backgrounds have been made.[16] The investigations concerned with legibility variations are almost entirely in agreement. The broadest investigation lists the following results from testing 850 subjects:

TABLE III[17]

COLOR COMBINATION	PERCENT SLOWER READING
Black on White	0.0
Green on White	3.0
Blue on White	3.4
Black on Yellow	3.8
Red on Yellow	4.8
Red on White	8.9
White on Black	10.5[18]
Green on Red	10.6
Orange on Black	13.5
Orange on White	20.9
Red on Green	39.5
Black on Purple	51.5

An interpretation of these and similar results of other tests is difficult because of the many variables over which no control was held or at least listed. Chief among these are chromatic aberration, the defining power of the various colors, their precise notation, and particularly their value (brightness) differences. Paterson and Tinker recognize the importance of the last when they state, "legibility depends . . . on the *brightness contrast* between print and background." They even feel that they are justified in calling it "the law of brightness contrast."[19] There is no question that value contrast is significant but Paterson and Tinker seem to have overlooked the fact that in another of their tests, held with precisely the same controls, white on black is reported at 10.5 percent slower than black on white. Of course, the value contrasts are the same. Obviously, then, other factors enter into the problem of legibility. There seems to be no doubt, however, that it is safe to make the generalization that dark print on a light background is more legible than the converse, and that as the value contrast decreases, so does the legibility.

Legibility is only one of the aspects of the lettering technique and perhaps not the most important at that. Certainly in situations where attention value is desired, perceptibility is of equal, if not greater, importance. Again, it is to be expected that the general rule of value contrast will hold and the evidence supports the contention.[20] There is some evidence, as might have been deduced, that colored lettering on a neutral or gray background was the most efficient.[21] It is a well known phenomenon that neutrals (gray) in juxtaposition with a color enhance the effectiveness of the color.

Lettering is an integral part of the cartographic technique rather than being extraneous to the fundamental purpose of the map. As the entire map is functional so should be the lettering. An examination of the problems of lettering makes evident the existence of certain basic principles. These principles are of two types: (1) those concerning the suitability of lettering for the intended purpose, and (2) those concerning the visual process of reading. That these cannot be separated is obvious.

Although the above enumerated general bases for the evaluation of the lettering technique apparently are sound, the data necessary for detailed objective evaluations are meager. Further research is clearly necessary, *aimed at the special requirements of cartography.* The designing of faces for cartographic use with particular reference to their reduction and orientation is perhaps most important. It is no accident that good freehand lettering is better than any other.

NOTES

1. J. Paul Goode. "To the Student and Teacher," *Goode's School Atlas* (Chicago, 1932), p. xiii.
2. The following table gives the estimated largest and smallest type sizes used in twelve well known atlases.

RANGE OF TYPE SIZES IN ATLASES

ATLAS	ESTIMATED AVERAGE SIZE OF SMALLEST LETTERING IN POINTS	ESTIMATED AVERAGE SIZE OF LARGEST LETTERING IN POINTS
a*	3–4	10–12
b	4	10–12
c	4–5	10–12
d	3–4	9–12
e	4	10–11
f	4–5	12–14
g	4–5	10–12
h	4–5	8–10
i	3–4	12–14
j	3–4	12–14
k	3–4	8–10
l	3–5	10–12

*a. Justus Perthes (H. Hack), *Steiler's Atlas of Modern Geography*, 10th Ed. (1925).

b. Touring Club Italiano, *Atlante Internazionale del Touring Club Italiano* (Milan, 1927).

c. Geogr. Anst. Von Velhagen u Klasing, (E. Ambrosius, Ed.), *Andrees Allgemeiner Handatlas*, 9th Ed. (Bielefeld and Leipzig, 1928).

d. J. G. Bartholomew and A. J. Herbertson (Alexander Buchan, Ed.), *Atlas of Meteorology* (Edinburgh, 1899).

e. George Philip and T. Swinborne Sheldrake, Eds., *Putnam's Economic Atlas* (London, 1925).

f. Topografische Dienst, *Atlas van Tropisch Nederland* (Batavia, 1938).

g. Edgar Ansel Mowrer and Marthe Rajchman, *Global War, an Atlas of World Strategy* (New York, 1942).

h. Erwin Raisz, *Atlas of Global Geography* (New York, 1944).

i. Rand McNally Co., *World Atlas* (Chicago, 1944).

j. Richard Edes Harrison, *Look at the World, The Fortune Atlas for World Strategy* (New York, 1944).

k. *Great Soviet World Atlas*, Vol. I (Moscow, 1937).

l. J. Paul Goode, *Goode's School Atlas*(Chicago, 1932).

3. Paterson and Tinker, *How to Make Type Readable*, pp. 29–37, 148.

4. Matthew Luckiesh and Frank K. Moss, "The Quantitative Relationship between Visibility and Type Size," *Jour. of the Franklin Inst.*, CCXXVII (1939), 87–97.
5. Ibid., p. 89.
6. Arthur H. Robinson, "The Size of Lettering for Maps and Charts," *Surveying and Mapping*, X, 1 (1950), 37–44.
7. Paterson and Tinker, p. 24.
8. See A. G. Eldredge, A. W. Abrams, W. Jansen and C. M. Shyrock, "Maps and Map Standards," *The Teaching of Geography* (Bloomington, Ill., 1933), p. 402.
9. W. D. Scott, *The Theory of Advertising* (Boston, 1903), pp. 138–139, and D. Starch, *Advertising* (Chicago, 1914), pp. 189–190.
10. Cf. Raisz, *Cartography*, p. 349; Jasper H. Stembridge, *The Oxford War Atlas* (London, 1941–1945), Vols. I–IV; Emil Herlin and Francis Brown, *The War in Maps* (New York, 1942). The development of film negatives for reproduction has made it relatively easy to reverse lettering on any background.
11. W. D. Scott, pp. 138–139.
12. Paterson and Tinker, p. 115; H. L. Hollongworth, *Advertising and Selling* (New York, 1920), pp. 76–78; A. Kirschmann, "Uber die Erkennbarkeit geometrischer Figuren und Schriftzeichen in indireckten Sehen," *Arch. f. d. ges. Psychol.*, XIII (1908), 352–388; and S. Slefrig, *The Normal School Hygiene* (London, 1905), Chap. 17.
13. G. Holmes, "The Relative Legibility of Black Print and White Print," *Jour. of Appl. Psych.*, XV (1931), 248–251.
14. C. D. Taylor, "The Relative Legibility of Black and White Print," *Jour. of Educ. Psych.*, XXV (1934), 561–578.
15. Starch, *Advertising*, pp. 189–190.
16. M. A. Tinker and D. G. Paterson, "Variations in Color of Print and Background," *Jour. of Appl. Psych.*, XV (1931), 471–479; M. Luckiesh, *Light and Color in Advertising and Merchandising* (New York, 1923), pp. 246–251; F. N. Stanton and H. E. Burtt, "The Influence of Surface and Tint of Paper on Speed of Reading," *Jour. of Appl. Psych.*, XIX (1935), 683–693; M. Luckiesh and F. K. Moss, "Visibility and Readability of Print on White and Tinted Papers," *Sight Saving Review*, VIII (1938), 123–134; K. Preston, H. P. Schwankl, M. A. Tinker, "The Effect of Variations of Color of Print and Background on Legibility," *Jour. of Gen. Psych.* VI (1932), 459–461.
17. Paterson and Tinker, *How to Make Type Readable*, p. 120.
18. Inserted from Paterson and Tinker, p. 113. Test covering 280 subjects.
19. Paterson and Tinker, p. 121.

20. Cf. Preston, Schwankl, and Tinker, "The Effect of Variations of Color of Print and Background on Legibility;" M. F. Miyake, J. W. Dunlap, and E. E. Cureton, "The Comparative Legibility of Black and Colored Numbers on Colored and Black Backgrounds," *Jour. of Gen. Psych.* III (1930), 340–343.

21. F. C. Sumner, "Influence of Color on Legibility of Copy," *Jour. of Appl. Psych.*, XVI (1932), 201–204.

VII

Map structure

A MAP is a visual composition and its planning is like that of a written paper. Like any written outline the plan for a map should be based on a reasonably complete outline of procedure, including the apportioning of relative emphasis to the elements of subject matter. In the preparation of written presentations the outline is concerned, in a sense, with relative position and size *in one dimension*. That is to say, the reader starts at the beginning and progresses in one direction to the end. Emphasis is primarily dependent upon arrangement, choice, and number of words. A visual composition is considerably more complex. The possibilities of arrangement are much greater because their positioning is always in terms of *two dimensions* and in many cases with at least an illusion or suggestion of a third. The techniques of visual emphasis, clarity, and order of viewing are made more complex, not only because of the two dimensional presentation, but because the possibilities in terms of technique combinations are almost unlimited.

The construction of a functional cartographic presentation according to plan is similar to the designing of any visual composition. Design in composition means order. Ross has summarized this accepted meaning of design in the following words:

> What we aim at is . . . a form of expression which will be simple, clear, reasonable, and consistent as well as true. The attention must be directed to what is important, away from what is unimportant. Objects, people, and things represented must be brought out and emphasized or suppressed and subordinated, according to the idea or truth which the artist wishes to express. The irrelevant must be eliminated. The inconsistent and the incongruous must be avoided.[1]

To follow such a procedure in visual composition is more exacting, while in a sense the control is less exact than it is in written forms of presentation. In almost all small-scale maps one of the basic "rubrics" of the outline, overall shape and position of land and water masses, is prescribed for the cartographer regardless of the specific function of the final presentation. For example, one area which presents a difficult organizational problem is Japan. To utilize a rectangular shape means the waste of at least 50 percent of the map area if Japan itself is the focus of interest. If the reader is not convinced let him try a map of Chile!

The visual elements within the overall structure of a map are numerous and are capable of almost infinite variety. Particularly is this the case with the so-called special purpose map which is unencumbered by style sheets and standard specifications. Writing more than thirty years ago, Hermann Wagner recognized the importance of design in the special purpose map when he wrote:

> In order to draw such a map, the cartographer must himself be intimately acquainted with the authentic material on which it is based; he must use the most careful discrimination in selecting and arranging this material for the purpose of the map; he must insert every single feature, contour, name, figure, with the thought in his mind of their probable effect on the unfinished map; he must, in the actual process of drawing, give account to himself whether and

why he ought to make a certain line lighter or heavier, a certain slope more or less steep, etc. He must, in one word, carry in his mind a perfect picture of that which he is to represent cartographically.[2]

From Wagner's characterization it is apparent that he felt that the special purpose map depends to a large extent on the cartographer's sense of design. Unfortunately nowhere in the literature of cartography is there any but passing treatment of the principles of visual design, except perhaps in connection with color. That there are principles of design applicable to the cartographic technique has been recognized in recent years but the recognition has not been accompanied by adequate review of the bases for its evaluation.[3]

Fundamental to the consideration of design in the cartographic technique (small-scale) are the problems and principles underlying the first step in any mapmaking, that is to say, the visual bases for choice of map projections. This involves the recognition of distortion of shape and area from the visual point of view, rather than from the mathematical, and the techniques for evaluating such distortion. The conventional use of projections, the relation between their intellectual and visual function, and finally the question of orientation are all significant rubrics in the evaluation of map design.

No visual technique is more important in cartography than that of contrast. Although it is difficult to generalize concerning the relative importance of a part of an integrated whole, it is likely, all things considered, that contrasts, of color, shape, size, and direction assume the dominant role in the majority of special purpose maps. The principles governing the use of shapes, line widths, and the other elements of contrast have been carefully applied to the fields of painting and advertising layout, but have had little recognition in cartography. Similarly the bases for evaluating the use of visual balance and proportion in cartography are poorly understood, if at all. Success so far has depended on the "mapmaker's taste and sense of harmony."

It is difficult indeed to segregate the various aspects of design for application to cartography but it is necessary because of the straightforward intellectual function of cartography. One of the most important techniques affecting the design of a map is the use of color. The problem of color in cartography is better left for discussion by itself because of its intellectual relationship, although it should be borne in mind that each of the principles of design included under the general headings of contrast and balance is markedly affected by the use of color.

Accordingly the utilization of the qualities of hue, intensity, and value of colors (and black and white) has been here separated from the other design elements, and it would not be strictly accurate to refer to the following aspects of the cartographic technique as being "design." In place of this general term which encompasses the entire composition of a visual presentation (including the lettering), the following may more properly be referred to as the structural elements of map design or simply as map structure, a term already in use in a more limited sense.

One of the many ways in which cartography differs from art is in the relative lack of flexibility in the modes of presentation on maps. This limiting factor in the visual design of a map makes itself felt in a number of ways, but none so fundamentally as in the matter of map projections.[4] The projection constitutes a systematic reference frame and once chosen does not allow arrangement of the shapes defined thereon except through changes in orientation. The initial choice of a projection, however, is important from the point of view of the visual structure of the map, since the projection system, especially on small-scale maps, determines to a considerable extent the basic arrangements of shapes. The possibilities of organizing shapes are greater than might be expected considering the precise nature of projections. There is a general tendency in the literature concerning projections to complain, or at least imply complaint, because it is impossible to represent truthfully the spherical surface on a plane surface. From the point of view of visual structure, the age-old problem of mapmakers might almost be considered a boon instead of an evil, since the opportunities for structural design would be greatly limited were the perfect map possible.

The subject of map projections has many facets, a number of which are of primary importance to the intellectual aspects of maps, but which have little or no importance with regard to the visual. The very purpose of many large and some small-scale maps prescribes, or greatly limits, the choice of projection in such cases as orthomorphism for navigation, true azimuths from the center for radial plotting, minimum error and adjacent fit for sheet survey, stereographic for circles, and some others. For such maps intellectual function is almost the sole basis for choice. A second basis or group of bases for the choice of projection is required in those cases where intellectual function and visual structure together form the foundation for choice. For example, distribution maps require equivalence, for obvious visual reasons,[5] which limits the choice to one group of projections, but structural considerations form the basis of selection therefrom. The combinations of intellectual function and visual structure which prescribe or greatly limit the choice of projection are many.[6] Besides the basic requirement of equivalence for distribution maps, many kinds of maps require specific combinations such as straight parallels and equivalence for presenting most climatic data, equivalence and minimum shape distortion for atlas maps, and minimum error and conformality for flight strip maps. In a great many cases, particularly with special purpose maps wherein the primary function is the presentation of a fact or a small group of relationships, the visual structure alone commonly assumes the significant role and should

The choice of a projection cannot, on the other hand, be made wholly dependent upon visual considerations, except possibly in the field of cartographic illustration when the illustrative function and the fixing of a general visual impression is of fundamental importance. The popular use of the oblique orthographic in recent years is a case in point.[7] There are, however, certain visual-intellectual relationships inherent in the earth grid, in addition to the well-known properties of map projections, which play an important part in the choice of a map projection for special purpose maps. The simplest and yet the most fundamental of these visual-intellectual relationships is the degree of harmony between the arrangement of lines on the sphere and on the projection. These are the most obvious of several visual characteristics of the earth grid:

1. All parallels are parallel.
2. The distance between parallels is (at small scales) equal.
3. If the eye faces squarely a point on the globe, the meridian of that point appears as a straight line.
4. All meridians converge or diverge.
5. Intersections of lines (except at the poles) are right angles.
6. All meridians are equally spaced along a parallel.
7. All meridians are great circles, the shortest distance between points. Hence, by visual implication, all meridians are straight lines.[8]

If the map reader either consciously or unconsciously knows these fundamental facts then it is safe to say that he will unconsciously accept any projection on which they are, or appear to be, reasonably presented. Evidence that he will do so exists in the lack of criticism of most conic projections and particularly of the orthographic. The latter is especially interesting since its sole attribute of significance is its visual quality. Except for this, it would be difficult to find a more useless projection since scale, directions (except from the center of the hemisphere), distances, shapes, and areas all vary considerably. It should be pointed out, however, that beyond the bare essentials of the grid system, visual comparisons from grid to grid are difficult because of the inherent complexity of the system. The eye is quick to notice departures from normal, and when in a map projection, any of the above list of visual characteristics is defied it is a source of irritation.[9] This irritation is considerably lessened by familiarity with the subject of projections. Nevertheless, the cartographer should attempt to reduce such discrepancies to the absolute minimum, particularly since the projection itself is not usually the point of interest in a special purpose map.

The visual-intellectual relationship to function of map projections is a subject which has not received much emphasis in technical writing. Many are the examples of intellectual absurdity in regard to choice of projections when due regard has not been paid to the visual aspects. The prime and well-known example is the long argument concerning the desirability of equivalence as opposed to orthomorphism, [10] and its reduction to absurdity in the case of the small-scale Albers versus the Lambert conic. The argument has real merit when large scale technical use is required, but Marschner summarizes the seemingly obvious answer for small-scale use when he points out that, "when used for the same zone with the same standard parallels, the Lambert conformal and the Albers equal-area projections are so much alike in appearance that, from a visual inspection, they appear identical." [11] An example of the degree with which the intellectual has sometimes been given precedence over the visual is the choice of the Minimum Error Rectified Conical Projection with two Standard Parallels for a map of the British Isles at a scale of 1:1,000,000 in place of the Albers equal-area. [12] The opposite, when the visual has been given entire precedence over the intellectual, is not uncommon in popular cartography. For example, many of the maps, even those of relatively large scale, in the Office of War Information Atlas, were thought to be made more visual by being constructed on the orthographic projection! [13] Obviously, between these two extremes there is a meeting place where the intellectual and visual aspects of projections can be brought to balance. All that is required is an analysis of the functional requirements and a fitting of the visual properties. It is doubtful if the search for the proper visual attributes necessary to satisfy any particular intellectual requirement need be unfruitful, considering the hundreds of projectional possibilities.

The literature on map projections, understandably, emphasizes the mathematical aspects, while their visual attributes and shortcomings have received relatively little attention. A number of systems have been devised to present visually the characteristics of different parts of a projection and the difference in distortion between projections. Reeves, [14] Jameson and Ormsby, [15] Fisher and

Miller[16] and Chamberlin,[17] among others, have plotted various figures ranging from a man's head to an equilateral triangle on different projections to serve as a visual means of comparison. The most systematic approach to analysis of deformation of shape and scale is, of course, the work of Tissot, who developed, in the latter part of the nineteenth century, his indicatrix based on the earlier work of Germain.[18]

The most fundamental visual problem, however, yct remains to be investigated. The restricted application of Tissot's analytical device "is of very little significance in considering the effect of distortion on the positional relationship between any two points at some distance from each other."[19] The only nongraphical approach is Hinks' statistical analysis showing error of distance and azimuth from the centers to the corners of some continental maps.[20] There does not seem to have been any other attempt to compare the relative magnitudes of the deformations on the various projections. Until such a study is made, the comparison of projections with regard to the character of the overall deformation must remain on the less satisfactory base of visual judgment.

One of the strongest conventions in cartography is the orientation of maps with north at the top. Happily, it seems to be losing some of its grip, but it is difficult to understand how such a convention could have been so strong for so long. It is a universal tendency recognized in both the art of composition and in the use of maps that the top of the rectangle is "farther away in the direction a person is facing." Notwithstanding the fact that for many centuries it was standard practice to assign orientation on some logical basis, even if it were only religious or national interest,[21] north orientation became fixed in cartographic methodology. It may well be that the convention had its origin in the visual qualities of the earth grid since orientation other than polar would create an unsymmetrical grid. At any rate, the convention is so well established that when a map is not oriented "properly" it is considered unorthodox.

From the point of view of education, north orientation has little justification. It has been defended on the grounds that it is fundamental to the teaching of shapes and areas.[22] Nothing could be more superficial since it is well recognized that shapes of objects are not actually learned when presented always in the same aspect. This is particularly true in regard to the locational relationships between shapes and areas. North orientation has even less justification in function since an experienced map user automatically turns the map so that "top" is in the direction he is facing.

It would seem logical for the orientation of a map to be based primarily on visual grounds which have their foundation in the intellectual function of the map. Thus the practice of route maps might be extended to all maps so that if there is a significant amount of "direction" in the function of a map it could be so oriented. Other visual aspects will, of course, enter, such as the incongruity of the placing of a large number of names at an angle to the grid, and the shape of the format. In general it is safe to say that many maps would benefit in terms of their inherent structural relationships if the convention of north orientation had not become established. Nevertheless, north orientation will, no doubt, continue to be common since in many instances the visual values of other orientation will be overbalanced by the resistance of the clientele to any disregard of conventions.

NOTES

1. Denman W. Ross, *A Theory of Pure Design* (Boston and New York, 1907), p. 7.
2. Hermann Wagner, "Zur Geschichte der Gothaer Kartographie," *Petermann's Mit.*, LVIII, 1 (1912), 14. Adapted by M. K. Genthe, "Notes on the History of Gotha Cartography," *Bull. of the Amer. Geog. Soc.*, XLV (1913), 34.
3. American Society for Professional Geographers. Committee on Cartography. "Cartography for Geographers," *The Professional Geographer*, IV (1946), 10–12.
4. Cf. F. J. Marschner, "Structural Properties of Medium and Small Scale Maps," *Annals Assn. of Amer. Geog.*, XXXIV (1944), 41. "But the extent to which medium- and small-scale maps retain the power of recording factual information in regard to distance, direction, and areal extent is first of all dependent on the structural property of the framework on which the map is drawn. The map projection controls position and expanse in the feature expression of the map."
5. See Irving Fisher and O. M. Miller, *World Maps and Globes* (New York, 1944), p. 107.

6. See J. A. Steers, *The Study of Map Projections*, 7th Ed. (London, 1949), pp. 234–252.
7. See the prefaces to Richard Edes Harrison, *Look at the World* (New York, 1944), and Erwin Raisz, *Atlas of Global Geography* (New York, 1944). In connection with the use of the orthographic, each points out that basic visual impressions of "global" spatial relations was considered the primary function of the "global" maps rather than the data thereon.
8. See Arthur H. Robinson, "An Analytical Key to Map Projections," *Annals Assn. Amer. Geog.* XXXIX (1949), 283–290, for discussion of the visual properties of the graticule.
9. Cf. S. [Reviewer], "The New Stieler Hand Atlas," *Bull. of the Amer. Geog. Soc.* XXXVI (1904), 266. The reviewer, writing of rectangular atlas maps cut from a large map and thereby capable of being joined, notes that "moreover, it causes the central meridian to run obliquely across the sheet, to the perpetual irritation of cartographic nerves."
10. Eckert, *Die Kartenwissenschaft*, I, 153 ff.
11. Marschner, *Structural Properties*, p. 29.
12. C. F. Close, *The Projection for the Map of the British Isles on a Scale of 1:1,000,000* (London, H. M. Stationery Office, 1903). (Albers was discarded because the map was not to be used as a base map!)
13. Office of War Information, *A War Atlas for America* (New York, 1944).
14. E. A. Reeves, *Maps and Map Making* (London, 1910).
15. A. H. Jameson and M. T. M. Ormsby, *Elementary Surveying and Map Projection* (London, 1942).
16. Fisher and Miller, *World Maps*.
17. Wellman Chamberlin, *The Round Earth on Flat Paper* (Washington, 1947).
18. Auguste Tissot, *Mémoire sur la Représentation des Surfaces et les Projections des Cartes Géographique* (Paris, 1881). The work was also published without tables, in Vols. XVII, XVIII, XIX (1878, 1879, 1880) of the 2nd Series of the *Nouvelles Annales de Mathematiques*. See also: Arthur H. Robinson, "The Use of Deformational Data in Evaluating Map Projections," *Annals Assn. of Amer. Geog.*, XLI (1951), 58–74.
19. Fisher and Miller, *World Maps*.
20. Arthur R. Hinks, *Map Projections*, 2nd Ed. (Cambridge, Cambridge Univ. Press, 1921), pp. 105–108.
21. Raisz, *Cartography*, p. 80.
22. F. P. Gulliver, "Orientation of Maps," *Bull. of the Amer. Geog. Soc.*, XL (1908), 538–542.

VIII

Map design

As pointed out earlier the cartographer does not have the same freedom of choice in his designing of a map as does the artist or advertising layout man, but nevertheless many opportunities are open for the adjustment of the elements of a map. One of the most fundamental of these opportunities lies in the visual structure of a map. In considering structural design it is well to bear in mind that all the design elements, like all the components of a map, are interdependent. They can be separated only for examination of their methodologic bases. Of the elements probably the one with the most widespread application is that of contrast. The components of a map which are capable of being varied according to the principles of contrast are (besides the lettering): the legend boxes, the title area, the graphic scale, the symbols used, and especially significant in special purpose maps, the various lines employed throughout the map.

In cartographic technique the problem continuously facing the cartographer is to assign visual importance and distinction commensurate with the intellectual significance of the ideas being presented. Such components as the legend box, title, and graphic scale, are not always of equal importance to the particular map or among similar types of maps. For example, the title of a map of North America may ordinarily be considered as relatively unimportant since the area so shown tells by its shape the coverage involved. On the other hand, the title on a map of a section of the continent or of an unfamiliar area such as a minor island is considerably more significant.[1] The legends of many maps are of little importance particularly when the symbols used are well-known or self explanatory. In other cases the legend box may hold the key which makes the map intelligible. In some maps scale relationships are fundamental whereas in others the scale is incidental. Some items symbolized are more significant than others. Similarly the various lines on a map vary in importance.

The principles of contrast include several phases, of which contrast of size is of fundamental significance. Contrast of size has two main points, all other aspects being equal:

1. The larger a visual item is, the more important it appears;
2. Equal size divides interest equally and produces confusion and monotony.[2]

Size contrast cannot be manipulated cartographically in the direct way it can when the designer has complete control over the shapes which enter into his composition. For example, no matter how significant the title of a map may be, it is doubtful, or at least very unlikely, that it can be made the largest item in the map structure. Similarly, the legend box and the graphic scale, although capable of variation cannot, in the majority of cases, be made to overshadow the map itself. Once the appropriate degree of intellectual emphasis has been assigned to each of the components, the principle of contrast of size may be applied by varying the sizes, within the limits possible, from the design that would result if all components were equal in importance.

Size contrast is particularly important with respect to the lines appearing on a map. Lines on the earth are generally predetermined in length, complexity, and overall shape by the existence of the phenomenon to be portrayed. The only opportunity the cartographer has to vary the lines is in the manner of their size, detail, and in character. A great many maps are apparently made on the assumption that the latter is most significant in creating contrast. Although no tests of relative visibility or degree of visual distinction between line weights and character are known, it would seem logical that character of line is not as great a factor in determining relative visibility as is size. All other things being equal the angle subtended at the eye by the object is the determining factor in visibility.[3] Whether a line is made up of dots and dashes, or any other combination, it is still a line which varies in thickness.[4]

Another manner in which contrast may be varied in cartography is that of shape variation. Although the basic shape with which the cartographer must deal is predetermined by the area, he is, within limits, able to vary the character of the shapes of such components as the legend box, blocks of lettering, symbols, and lines, either in the legend or elsewhere on the map. The principles of shape contrast which apply to these components of a map are similar to those of size contrast, namely, that monotony or extreme complexity of shape retards the visibility, while simple differences in shape promote interest and enhance visibility.[5]

So far as is known no tests have been made to determine the quantitative visibility characteristics of the basic shapes. Undoubtedly there are varying magnitudes of dissimilarity of shape which increase visibility. Conversely there are undoubtedly values of equality of size, given different shapes, which are significant if the cartographer wishes to distinguish between two or more components, but must assign equal visual values to each. It is common knowledge, for example, that a round shape seems larger than a square shape with the same area.[6] More exact understanding is needed.

Another aspect of shape contrast of considerable significance to the cartographer is that, other things being equal, the more complex a shape is the greater interest it seems to have for the eye. The operation of this principle is particularly noticeable on many maps in the relation between the coastline and the linear data thereon. It has been pointed out that accurate intellectual generalization "should mean the same degree of omission of detail for each kind of feature."[7] Such generalization often results in natural features being remarkably complex visual shapes and thereby attracting the eye away from the more important elements of the presentation.[8] Raisz recognizes the advantages of visual simplification and has noted that there is often "a discrepancy between the details of the shoreline and the broad generalized belts placed upon them."[9] Although intellectual simplification or generalization is of fundamental importance in mapmaking the problem of *visual* generalization, in relation to map structure, although not so well understood, is of almost equal importance to the clear presentation of scientific data.

The third manner by which the importance of contrast manifests itself in visual structure is the phenomenon of eye movement that results from variations in the shapes, values, and lines. Like all elements of visual design it is difficult to separate this aspect from the other elements.

Ross, in his detailed analysis of the principles of design, shows many ways in which the eye can be induced to move (called "direction" by the artist) through the manipulation of positions, lines, outlines, and values.[10] These techniques are important in considering the structure of a map since, like most visual compositions, the data of a map cannot be grasped in just one glance or by concentrating on only one point. The eye, in order to encompass the entire presentation, must move from place to place. Since the components of the map are obviously of differing importance, it would be desirable to lead the eye in the direction and sequence necessary for the proper grasp of the complicated material. In other words, the principles of movement would help the cartographer to construct the presentation in visual accordance with the intellectual aims.

The first principle of movement in visual structure is that "every unit in the design of a layout, unless it is of the same width as height, points in the direction of its longest dimension."[11] This principle, of considerable importance in advertising, has less application in cartography since the fundamental shape (the area mapped) is already determined; whereas in advertising there is danger of the reader being led off the layout to an adjacent one. Since we can assume the map reader wants to look at the map the above principle is important in directing his eye toward the significant features thereof, which may be the legend, scale, title, particular area or relationship, or marginal information. De Lopatecki also suggests that "any unit that has a definite pointed shape will point,"[12] and suggests that the general rule for stopping eye movement is to arrange another unit at right angles to the first.[13] Although no tests or experiments are known to exist concerning the application of these principles it is likely that they are significant in cartography not only in the arrangement of map components, but in the overall map shape as well. It would be instructive to have the results of visual tests conducted on the relative efficiency with which concentration was induced by the oval world projections, such as Mollweide and Aitoff, compared to the interrupted pointed projections such as the homolosine and parabolic.

The various factors of contrast often are found occurring together especially in what is sometimes called the figure-ground relationship. This aspect of design refers to the visual relation of one or more components to the background on which they are seen. Although it becomes most apparent in symbol shapes and in maps suggesting pictorially the third dimension, it is also of considerable importance in helping to define the area of interest in a map. For example, the use of water lining, shading stipple, or perspective raising of coastlines are devices which make clear the land-water relationships.

One of its most obvious manifestations is in the use of spherical or cubical (block pile) symbols, the shape of which is intended to raise them above the general map level.[14]

One of the essentials of writing deals with the maintenance of a proper relationship among the features being presented. The appropriate amount of detail and emphasis should be applied to each element. A visual composition likewise ought to be devised and arranged, insofar as possible, so that each component appears logical both as to position and degree of emphasis. There is little fundamental difference between the construction of a written outline and a visual outline. The techniques are different, but the functional bases are the same.

In the cartographic technique the visual "outline" is a complicated combination of the application of the principles of overall shape, balance, proportion, and unity. These general principles, difficult to define and poorly understood scientifically, enter into every visual composition. More directly than in the case of writing, each element modifies the other and an almost infinite variety of arrangements is possible. Many of the principles have been derived by empirical methods through many centuries of work by artists and designers. So far as is known the principles have not been tested in specific application except insofar as their persistence against competition may be considered testing. Until such time as logic and objective research concerning the relative efficiency of the various possibilities is undertaken, the cartographer can but rely on the experience and direction of the artist.

Cartographic method as concerned with map structure includes the problem of format or overall silhouette, as distinguished from the shapes entering into structural relationship within the map. Ordinarily the cartographer, like the advertising layout man, must work with a prescribed format but within that shape he is usually able to exercise some freedom. Although the literature is singularly bare of references to this problem the numerous possible approaches indicate that some basis for judgment would be desirable. Eckert attempts to generalize on a visual-intellectual basis by suggesting that the format should be rectangular and only circular or oval when it concerns the whole earth.[15] There has recently been a tendency to return to the circular shape in the "air age maps," presumably to emphasize the sphericity of the earth.[16] A number of world maps are "interrupted," such as the homolosine and show quite unique and irregular outlines. The problems and cartographical results of arranging desired scale, area shape, and prescribed format are legion.

Fundamental to the problem of format is the principle of unity. De Lopatecki summarizes the principles of unity by pointing out that "all the units which go to make up a layout must be organized into a single whole,[17] although he was not considering a format as being involved since there is a strong tradition in advertising layout concerning format. Keeping in mind the principle of unity it may be seen that the elements of structure outlined under contrast, particularly concerning directional components, are of considerable importance. For example, the pointed shapes of the interrupted sinusoidal tend to direct the eye away from the map, and do not in any way help to suggest the desirable unity.[18] On the other hand if the same or a similar projection is used with a border,[19] or with such strong color contrast as to center the attention,[20] the principle of unity is not seriously violated.

There are a few visual procedures which aid the cartographer in working out format problems. The layout man usually works from a format to his composition while the cartographer, because of the prescribed shapes, must start with some of the more important shape elements already determined and, in many cases, their basic organization prescribed. The drawing of a frame or the decision concerning its scope can be greatly facilitated by bearing in mind the principles of balance and proportion. The principle of balance is well stated by De Lopatecki:

> Every layout requires stable equilibrium or balance. The masses must be so weighed against one another that they appear to have *settled* in the positions they occupy—to belong there in other words. No unit of design should convey to the eye of the reader the idea that it is struggling to go somewhere else in the layout, nor should the layout look as if it were tipping over.[21]

Visual balance begins with the focal center which lies on the vertical center line approximately 5 percent above the mathematical center. Around this optical center the units of a composition should be balanced visually. The factors affecting the visual balance of a composition are the distances from the center of the components, their weights (visual), their sizes, and their shapes.[22] In cartography, as previously pointed out, basic shapes and the intellectual importance of the components are usually predetermined and the cartographer must therefore analyze his subject matter, decide where he wants his optical center to be, and then construct his frame.

The principles of visual proportion as related to cartographic methodology may be summarized as being the relation which the size of one part bears to the whole and to the other parts.[23] In design the subject is of great moment and a large literature has been developed on the theories and history of proportion. Much of it is very technical, being based on mathematical and geometrical principles; but little is directly applicable in cartography since the cartographer does not have the necessary freedom to vary his shapes and sizes.[24] It does have application in the determination of the map frame, and in the shapes and positioning of the nonmap components such as legend-boxes, blocks of explanatory text and other shapes with which the cartographer is relatively free to work.

One of the fundamental principles of proportion seems to be that the rectangle is the most pleasing shape, although the circle has been described as the "most harmonious of all outlines.[25] The rectangle is the more common and because of its acceptance by designers and its conventional employment it may be considered desirable except when some other shape is clearly indicated. In layout and design considerable attention is paid to the actual dimensions of the rectangle, and elaborate schemes involving diagonals and reciprocals are employed in working out the internal structure. In cartography the area and scale usually dictate to a considerable degree the proportions possible; however, one of the negative applications of the principles of proportion should be employed in the choice of a rectangle. The square and the rectangle with obvious mathematical relationships between the sides is not considered as pleasing a figure as one wherein such relationships are not readily apparent to the eye.[26]

It is undoubtedly all too apparent from the foregoing that structural elements in the cartographic technique are not only extremely complex, but poorly understood as well. The reason that visual composition is so subjective and devoid of objective testing is probably, or at least partially, due to the assumption that, because of the infinite number of possibilities, any testing of isolated components would be of little actual worth. It seems likely, however, that a number of cartographic procedures could be evaluated by testing.

For example, there is undoubtedly a minimum width difference and angular difference easily apparent to the eye. It should be possible by testing to arrive at a reasonably accurate area departure factor which when applied to different shapes would bring them to comparable visual size. On the other hand, many of the aspects of harmony, movement, balance, and proportion, seem likely forever to remain essentially subjective insofar as their evaluation is concerned. This does not mean to imply that the principles governing their use are purely a matter of individual caprice; it does mean that *exact* standards probably cannot be devised.

The one aspect of map structure which seems to have definite possibilities for objective visual evaluation is the projectional. Heretofore, almost the entire literature on projections has been concerned with the mathematical phases. Only recently has the visual problem become significant in the selection of projections, but so far little has appeared in the way of logical argument beyond the self-evident assertions that a circle represents the hemisphere and that a perspective drawing of the surface of a globe looks like the surface of a globe. It is in the field of visual evaluation of projections having significant properties for small-scale cartography that investigation is required. Visual illustration of distortion has been presented in numerous cases but objective analysis is lacking. Tissot's investigations seem to provide the basis for such analysis. Considering the present state of cartography in regard to structure, it seems wise to suggest that design has a negative side as well as a positive. Excessive repetition, ignorance of figure-ground relationship, the use of uninteresting shapes, the lack of focus or contrast may well detract from the effectiveness of a map just as much as their opposites would add to its effectiveness.

NOTES

1. Cf. The formula developed by Fretwurst (Eckert, *Die Kartenwissenschaft*, I, 343) for title lettering! The formula reads, $h = 2.1 X\sqrt{I}$ where h equals the height of the lettering in mm., and I equals the area of the map within the neat lines.
2. De Lopatecki, *Advertising Layout and Typography* (New York, 1935), pp. 16–18.

3. Luckiesh and Moss, "The Quantitative Relationship Between Visibility and Type Size," *Journal of the Franklin Inst.*, CCXXVII (1939), 87–97.
4. It is reasonable to suppose that if the separate symbols, such as dots, used to simulate a line are much farther apart than their width that the eye will fix upon the individual pieces, so to speak, and the resulting visual impression will fundamentally be one of a series of jumps even though eye movement is induced by their arrangement.
5. De Lopatecki, pp. 18–20.
6. Ibid., p. 30.
7. John L. Ridgway, *Scientific Illustration* (Stanford, Stanford Univ. Press, 1938), p. 79.
8. A particularly good example are the two maps in Fig. 151 of southern South America in Clarence F. Jones, *South America* (New York, 1930), p. 345. Although side by side, and intended to show historical sequence (sugar and vineyards in Argentina and Chile) one map has a much more complex coast of Chile than the other.
9. Erwin Raisz, "Draw Your Own Blackboard Maps," *Jour. of Geog.*, XLI (1942), 236. Cf. also Eckert-Greifendorff, *Kartographie*, pp. 79–83.
10. Ross, *A Theory of Pure Design*, pp. 25–32, 79–130, 182–184.
11. De Lopatecki, p. 22.
12. Ibid., p. 24.
13. Ibid., p. 25.
14. Guy-Harold Smith, "A Population Map of Ohio for 1920," *Geog. Rev.*, XVIII (1928), 422–427 (Plate IV); Raisz, *General Cartography*, pp. 237–239.
15. Eckert, *Die Kartenwissenschaft* II, 680–682.
16. Erwin Raisz, "Map Projections and the Global War," *The Teaching Scientist*, II (1942), 33–39.
17. De Lopatecki, p. 49.
18. Cf. Army Service Forces, *Atlas of World Maps, M-101* (Washington, 1943), General Reference Map, p. 1.
19. Cf. *ibid.*, Landforms, p. 2.
20. Cf. Encyclopaedia Britannica, *World Atlas* (Chicago, 1945), pp. 3M–3N.
21. De Lopatecki, p. 26.
22. Ibid., p. 27.
23. Ibid., p. 33.
24. A brief summary is given in De Lopatecki, pp. 33–44. The theories of proportion and symmetry are of interest to the cartographer since it aids him in the appreciation and analysis of associated visual design.
25. Ross, *A Theory of Pure Design*, p. 98.
26. De Lopatecki, p. 33.

IX

Color in cartography

OF ALL the media of cartography, color is probably the most complicated and the least understood. Notwithstanding, it is a subject on which most cartographers and writers do not hesitate to discourse, for color is widely used in cartography and it is likely that its use has been more influential than any other technique in the creation of the conception of cartography as a branch of subjective art. This is not at all surprising for color has been the main medium of art throughout recorded history, and color has always stood a little apart from other representation techniques because of its closer alliance with things aesthetic. It is interesting to note that in the time of J. Perthes (*ca.* 1800), before color printing, the illumination of political boundaries and coastlines was done by the society ladies of Gotha strictly for enjoyment! [1]

Although color has always been considered important in cartography it did not come into widespread use until chromolithography was developed during the first part of the nineteenth century, and since that time color has come to be indispensable in important cartography. It is now considered so important that the making of a map without color is a kind of negative undertaking, that is to say, a black and white map is usually constructed because color could not be used. Eckert asserts: "For many of our modern maps, particularly the special purpose maps, colors are indispensable . . . black and white maps grow monotonous.[2] Those words were written, however, before the developments which make possible the tremendous variety of techniques for black and white presentation.[3] The procedures for widening the techniques of black and white work are not even now well known or widely used in cartography, and even if they were it is doubtful whether the general attitude toward the importance of color would be markedly affected.

In cartographic methodology the use of even one color in addition to black seems to increase enormously the clarity and the ease with which an item or group of data can be emphasized. This is frequently caused by the phenomenon noted earlier, namely, that the visibility of line work (lettering, grid, hydrography, etc.) is markedly lowered by having a colored background. Much of the enhancing of visibility through the use of color is likely to be rather accidental, in a sense, because of the rather sterile design common in black and white cartography, but even if the potentialities of design were widely employed, color would still assume a most significant place in cartographic methodology.[4] Color seems to produce effects beyond anything as simple as legibility, in the aesthetic sense as well as the nonaesthetic subjective. For some reason, as soon as color is applied to a map the color becomes a focus of criticism, and everyone, regardless of his familiarity with the principles of color use, seems to feel that he is entitled to comment upon the color on the basis of his own likes and dislikes. It is noteworthy that, although structural design principles and effects are frequently slighted in reviews, the use of color commonly occasions comment.

The increased color reproduction facilities as well as the growth of propaganda maps during the past decade or so, has brought a realization of the danger of using color haphazardly. Nearly forty years ago Eckert pointed out that "an artistic appearance, particularly a pleasing colouring, can deceive in regard to the scientific accuracy of a map.[5] During the recent war the propaganda uses of color were commented upon, and although the dangers were fully realized and presented, little seems to have been suggested to raise the standards.[6] It is surprising that this subject, acknowledged by most writers to be of extreme importance in cartography, has not been investigated to the fullest extent by cartographers. As a matter of fact the use of color in cartographic design received more study during the last half of the nineteenth century (beginning in 1850 when facilities for its use began to become available) than it has during the first half of the twentieth century.[7]

Prior to and even during the first half of the nineteenth century the color which appeared on maps was applied either by engraving (wood or copper), or by hand illumination. Even after the development of lithography, copper engraving, although more expensive, was still used for the more precise and valuable maps. It was not until the development of the various combinations of photography and metals with lithography, toward the end of the nineteenth century, that color became relatively common in cartography. By that time considerable investigation in the principles and characteristics of color vision had been made, and Peucker utilized these developments in his study of shadow plastic and color plastic in relief representation.[8] Following the early investigations of Peucker many others carried further the research into the employment of colors on maps, particularly in the field of landform representation. Considerable progress was made and the concepts of warm and cold colors, stereoscopic effects of colors, hue, value, and intensity differences were apparently fairly well known among the European cartographers. This period, prior to the First World War, may be characterized as the most productive in terms of the application of color science to cartography.

The progress in reproduction techniques after the First World War was rapid and although great strides were made in the technical procedures little or nothing was done to carry forward the research which more or less started with Peucker's efforts. As a matter of fact the period was in a sense negative rather than positive. While little if anything was done to further this line of research, a great deal that had been accomplished was forgotten, at least in American cartography. This is understandable, even if unfortunate. As the techniques of reproduction have grown more numerous and complicated the familiarity of the cartographer with such processes as deep etch, process color, wash, and the various screening techniques grew relatively less. As a result the cartographer, all too often not being competent to direct, left much of the color production responsibility to the lithographer and engraver. Consequently, the application of color to many maps was under the direction of those who were not only unfamiliar with the cartographic technique but were even less familiar with the presented data. It is not surprising that cartographic quality declined.

That the pendulum has begun to swing in the other direction, however slightly, is evident from a number of publications. *The 32nd Yearbook of the National Society for the Study of Education* asked for more research in the legibility and psychological effect of colors on maps. The *Encyclopaedia Britannica* employed a color artist for a series of world maps in its atlas. The experience of cartographers during the recent war resulted in a plea for more care in the selection of colors.[9] A number of government agencies have been investigating the problem of color. Unfortunately, the evidence of progress is meager and, although it indicates a move in the right direction, does not promise as much as would be desirable. No modern studies comparable to those of the time of Peucker have been undertaken and even the current critiques of maps show little familiarity with color science. Color work is described in meaningless terms as "pleasing," "of good quality," or "mediocre," with little apparent substance back of the comments. In the current literature one rarely sees even such comments as "Dr. Weise is not one of those geographic sinners who use any color for any step in a graded scheme of intensities."[10]

The loose criticism and the neglect of substantial color work in cartography would not be so serious if maps were actually problems in art, although as aesthetic art their quality would generally be low indeed. Color work in cartography becomes of extreme importance, however, because of its very significant physiological and psychological effects. That erroneous effects can easily be produced by deliberate color misuse has been clearly shown by Speier and Wright. Whether misuse is deliberate or merely the result of ignorance, the effects are equally unfortunate. It is no accident that many German wall maps prior to the war, and German propaganda maps during the war were better suited to their purpose than similar American products. The science of color vision has greatly expanded since 1900 and the physiological, physical, and psychological determinations of that period provide a relatively sound basis for intelligent color in cartography. Admittedly there is much, particularly in the psychological field, yet to be investigated, but it is safe to assume they are but refinements of the principles already known.

Another of the fundamental problems of color in use which has plagued the scientific worker is the difficulty of color description or terminology. Until recently there has been no method of describing colors which reasonably matched the physicist's descriptions of dominant wave length, purity, and brightness. However the Munsell, Ostwald, and Birren systems of color notation although not perfect provide the artist, cartographer, designer, and the general user of color with systematic and workable methods of color terminology. The Munsell system [11] has recently been adopted by the Bureau of Plant Industry to describe soil colors. [12] With it one may describe precisely the value and the intensity of a color although the terms for hue description are less exact.

Paralleling the development of the science of color vision, the techniques of color reproduction and color application have made remarkable strides in recent years. Probably no single technique has given greater potentialities to the field of cartography than the air brush. This technique, barely fifty years old counting the first crude mechanisms of the early 1890s, allows the smooth application of continuous tones by the cartographer on copy, an operation formerly difficult except for the highly skilled. In the field of lithography and other reproduction processes advances in technical procedure have been equally rapid. At the present time it is possible to reproduce practically any kind of cartographic copy. Unfortunately, as the techniques and scientific knowledge have increased, so has the cost, and at the present time costs are extremely high. This is due in part to production costs, but is likely more influenced by the lack of competition in the greatly understaffed and underequipped printing industry. As a result of the costliness of color production it is to be expected that the techniques for black and white cartography will be more closely investigated and better utilized. There will no doubt be, however, little decline in costs of commercial color reproduction and, although the majority of cartographers will not have the opportunity to assist in such undertakings, they should be familiar with the bases for evaluating the techniques used. Familiarity with principles of color use will also enable the cartographer to make an intelligent choice of the available alternatives when economic considerations preclude the use of the ideal. For example, a range of shades or tints may be produced from a single plate whereas the best range would involve the use of several plates. The choice of which particular hue to use then depends on the characteristics of hue perception and the purpose of the map.

The evaluation of color methodology must obviously be based on the characteristics of color vision. If one were interested in the production of color he would necessarily need to be familiar with the physical and chemical laws relating to color as light and pigment. A reasonable knowledge of these matters is necessary to the understanding of color vision, but:

> In addition to this the designer and user of color must be able to anticipate the often surprising action of color as the eye sees it, which is loosely called "the psychology of color." He must realize that a color which satisfies the chemist's and physicist's standards may not be satisfactory as the eye sees it either for the purpose for which it must be used, or the manner in which it is to be used.[13]

When color is examined from the point of view of its effect on the observer it has several characteristics. To the eye, color varies as to hue. This is analogous to the dominant wave length of the physicist. Because of the structure of the human eye and the differing wave lengths of light, human vision reacts in different ways to different spectral hues. A knowledge of these differences is quite obviously necessary to the proper employment of colors. Secondly, each individual hue varies both in terms of its inherent brightness and in the degree of saturation or intensity of the color area. Just as the eye reacts differently to the various hues so does it have varying sensitivity to value and intensity changes. Moreover, color always appears in an environment, and the environment has a marked influence on its appearance. A significant although less important aspect of color methodology concerns the conventions, preferences, and the traditional significance of colors. The cartographer must be familiar with all these considerations before he can effectively evaluate the color technique. He should also have some background in the basic elements of color vision and color science.

Color is the visual sensation produced by certain wave lengths of light. As is well known, the visible portion of the spectrum is relatively small and ranges from the short-wave violet to the long-wave red. Between these two extremes occur all the pure hues, in combination producing white, but excluding black, which is the absence of light. Practically all hues are combinations of various wave lengths. A pure spectral hue is rarely seen.

It is fundamental to the consideration of color that it be clearly understood that for practical purposes color exists only in the eye of the observer. The physics of light is of importance in the investigation of the characteristics of color behavior and its findings provide a solid foundation for the discussion and analysis of color perception. But the study of color whether in the cartographic technique or in any other aspect of its use, is based fundamentally not on the physics of light but on the sensations produced by the eye's reaction to colors.

One of the greatest difficulties connected with the study of color as a sensation is found in the description of colors. All who have tried to describe colors have experienced great difficulty in finding adequate words to express themselves, until Munsell and Ostwald placed color terminology on a consistent basis.

Colors vary as eye sensations according to hue, value, and intensity. Hue is the color itself, for example a red or a green. Value is the intrinsic lightness or darkness of a color, as for example, red is usually darker than yellow. The value scale ranges from black to white. Intensity is the term applied to the relative purity and amount of the color in a given area. This is expressed as a measure of the purity of the color compared to a neutral gray of the same value as the most intense expression of the hue under consideration. Thus a full scale of intensities would range from the hue on one end to a neutral gray on the other. At no place would it vary in value. The scale of value is the simplest of the scales and is usually considered as a scale of ten divisions with black being the lowest in value to white at the highest. The intensity scale is more complicated. In the Munsell notation system the most intense red capable of being produced was considered as being ten steps away from neutral gray of the same value. This series established the intensity scale. It is interesting to note that since Munsell established this scale the ability to produce more brilliant colors has necessitated the extension of the intensity scale at least four steps beyond the original ten. [14]

The description of hues is by far the most complex aspect of color notation because the possibilities of color combinations are infinite and because of the conflict of the physical and psychological methods. The physicist describes a color in terms of its dominant wave length, the extent to which it stimulates the red, green, and blue sensitive nerves, its brightness and its purity (intensity). Unfortunately the comparison and matching of colors and most other color sensations are not systematically stimulated by colors according to wave length, etc., and it is necessary in the use of color that a system of color description be devised based on visual experience. Working on this basis Munsell chose the four basic hues—red, yellow, green, and blue—and together with purple, arranged them around a circle. Between each pair of the above colors he placed the intermediate hues thus giving a hue circuit in which each hue is diametrically opposite its complement. Each hue is varied in terms of brightness in one dimension and intensity in the other.

Although Munsell's notation system has received wide acclaim in the United States it has been improved upon by Ostwald [15] and Birren [16] who have shown that, from the point of the user of color,

Munsell's system does not supply adequate understanding of harmony relationships. [17] First Ostwald and later Birren showed that all colors seen by the eye were derivatives of pure hue, black and white, and were susceptible of being given meaningful equations. These developments are of particular importance in providing the basis for color harmony and should be of great interest to the cartographer. It is impossible in this study to enter into the laws and principles of harmony, *per se.* As pointed out earlier the cartographic technique is less concerned with the creation of aesthetically pleasing compositions than it is with the problems of clarity and design, in the sense of appropriate representation. For the purpose of examining the bases for evaluating colors for cartographic use it is sufficient that it be established that colors vary in terms of hue, value, and intensity. Ostwald showed, however, that from the visual point of view "white and black were colors just as much as were red and green." [18] This is fundamental to the understanding of color methodology and is one of the most significant aspects wherein Eckert and the other writers on color in cartography failed in their efforts to establish sound bases. As a matter of fact, the earlier writers and users of color in cartography had to depend for their logic primarily on the research of the physicist, less on that of the physiologist, and hardly at all on that of the psychologist. [19] The realization that color in use should be considered primarily as a sensation came only later. Thus the early emergence of the convention of representing elevation according to the arrangement of the spectrum now lacks foundation in visual logic.

The fact that the eye distinguishes colors in terms of hue, value, and intensity by no means completes those aspects of color vision of importance to cartography. It has long been known that although a color may be physically determinable by reference to spectrophotometry or to a color tree or color solid, its visual character depends upon its environment. The well known phenomenon of simultaneous contrast occurring when dark and light areas are juxtaposed is a simple illustration. (See illustration facing p. 94.) The behavior of color vision with respect to environment is of considerable importance in the logical construction of maps.

NOTES

1. Eckert, *Die Kartenwissenschaft*, II, p. 688.
2. Ibid., p. 689.
3. Cf. Clarence P. Hornung, "Art Techniques and Treatments," *Seventh Annual Advertising and Publishing Production Yearbook* (New York, 1941), pp. 18–33. Over 100 possibilities are shown out of the "limitless number of variations possible."
4. Cf. Eckert-Greifendorff, *Kartographie*, p. 33.
5. Eckert, "On the Nature of Maps and Map Logic," *Bull. of the Amer. Geog. Soc.*, XL (1908), 347.
6. Cf. Hans Speier, "Magic Geography," and John K. Wright, "Map Makers Are Human."
7. The classic work on color contrast and harmony (M. E. Chevreul, *The Laws of Contrast of Colour, and Their Application to the Arts*, trans. by John Spanton, London), had "Map Coloring" as part of its subtitle in the original edition published in 1835. Chevreul's work has been ignored in cartographic literature.
8. A good summary of this work and its effects is contained in Eckert, *Die Kartenwissenschaft*, I, 625 ff.
9. American Society for Professional Geographers. Committee on Cartography. "Cartography for Geographers," *The Professional Geographer*, IV (1946), 10–12.
10. Mark Jefferson, "A New Density of Population Map of Europe" (A Review), *Bull. of the Amer. Geog. Soc.*, XLV (1913), 669.
11. Albert H. Munsell, *A Color Notation* (Boston, 1916).
12. "Report of the Committee on Soil Color to the 1946 Soil Survey Staff Conference," *Soil Survey Field Letter*, No. 2 (1946), pp. 13–14.
13. The International Printing Ink Corp. Research Laboratories. "Color in Use," Monograph No. 3 (New York, 1935), p. 5.
14. Ibid., p. 10.
15. Wilhelm Ostwald, *Colour Science*, trans. by J. Scott Taylor, 2 Vols. (London, 1931–1933).
16. Faber Birren, *Color Dimensions* (Chicago, 1934).
17. For a more searching analysis of the various color notation systems and their bases in color sensation see Faber Birren, *The Story of Color* (Westport, Conn., 1941), pp. 237–256.
18. Ibid., p. 255.
19. Cf. Faber Birren, "A Chart to Illustrate the Fundamental Discrepancies of Color Study in Physics, Art, and Psychology," *Psych. Rev.*, XXXVII (1930), 271–273.

X

The employment of color

THERE seem to prevail among otherwise logical thinkers the curious notions that a) what the individual himself likes another will also, and b) if someone likes a technique it is therefore good. If ever means and ends were confused it is in these reactions to color use. A close analogy can be made regarding food. The gourmet likes something because it tickles his palate; the nutritional expert tells us what promotes health. The two, unfortunately, frequently are not the same. In cartography our choice of color may be to please or it may be to promote the purpose of the map. Except within the broadest limits personal preferences should play a small role in the choice of colors.

Colors vary in a number of ways as we have seen. The reactions of the human mind and eye to these variations is obviously the primary basis upon which the color scheme for a map should be chosen. The most productive way, then, to approach the problem is by means of an analysis of the sensitivity of the eye to the ways in which colors may be varied. As in the other aspects of the cartographic technique, much yet remains to be accomplished before we will be able to make our choice with certainty. The researchers in color have, however, given us a considerable amount of data; to heed it would be to our advantage. The first group of these data concerns hue sensitivity.

It is difficult to generalize concerning hue sensitivity because the investigations so far deal primarily with sensitivity of the eye to purely spectral hues and largely disregard the effects of hue combination, the more common experience. The results of various experiments regarding spectral sensitivity reported by Luckiesh[1] are not entirely in agreement, but they strongly indicate that the eye is not ordinarily sensitive to changes of less than approximately ten millimicrons. This would mean that there would be less than thirty spectral hues normally identifiable by the average eye. It is likely that the average untrained person can see considerably fewer. Notwithstanding that such data are of great interest because they indicate that the untrained eye is not particularly sensitive to changes in spectral hues, the available data do not provide bases for judging the sensitivity of the eye to combinations of hues, the most common hue sensations. It should be possible, using spectrophotometric methods, to arrive at a distinguishable interval for variations in hue.[2] Birren points out that, while probably at the very maximum not more than 900 or 1,000 may be seen by the eye, our vocabulary for color is very small.[3] In general the available data would seem to indicate that choice of colors for cartography should be made with full realization that the untrained eye does not have much ability to distinguish between colors, and that the farther hues can be separated (without destroying any required unity) the better.

The eye is definitely more sensitive to some hues than to others. All observers agree that the eye is most sensitive to red, followed by green, yellow, blue, and purple, in that order.[4] This series provides the cartographer with a partial basis for choice of color depending upon how much emphasis is desired for the data to be represented by a color. * Unfortunately no satisfactory data seem to exist which

* It should be borne in mind constantly that with respect to color, as with all visual phenomena, all aspects are interrelated so that no single physiological or psychological phenomenon provides the sole basis for choice. This fundamental principle should precede every discussion.

would make it possible to grade the colors according to relative degree of sensitivity.[5] A study by Féré commented upon by Luckiesh indicates that red has nearly twice the effect of blue on muscular activity.

In the study of the sensitivity of the eye to hues, it is well to remember that hues vary in terms of value. This variation, in turn, is of considerable significance with respect to the visibility of data because of contrast relationships. The relative luminosity of the spectrum for the normal eye has been determined. It is highest in the yellow-green region (555 millimicrons), and falls to approximately half its maximum within a range of fifty each way.[6] Disregard of this obvious relationship through blind subservience to the "logic" of the progression of the spectrum has led to many incongruities in mapmaking. Although the defect in the "logic" of the spectrum is primarily due to value differences, it is pointed out here as a good example of the interdependence of the various aspects of color in use. In many atlases, in the international map, on many small wall maps, and in many other cases where colors have been chosen for hypsometric shading of altitude, the spectrum has been used as the basis. The lower altitudes have been shown in greens, followed by yellows for the intermediate altitudes and then reds or near reds for the higher. Because of the relative luminosity of the spectrum, by far the lightest areas, and by comparison the most visible, are therefore the areas of intermediate altitude which are rarely the areas of great importance. The indiscriminate use of yellow with its high luminosity and its unfortunate effect in juxtaposition with other colors is well shown in the series of special maps in the Encyclopaedia Britannica Atlas.[7]

Of minor interest in cartographic methodology is the phenomenon of the shift of hue sensitivity with changes in illumination. With a decrease of illumination the maximum of the luminosity curve shifts toward the shorter wave lengths. In other words if a red and blue area are of equal brightness at a high illumination, the blue surface will be brighter at a lower illumination. Consequently it is obvious that maps should be constructed, in so far as possible, for the illumination in which they will be used.

It is a well-known fact that some colors appear more individual than others.[8] For example, red is red but orange seems to be composed of both red and yellow. Many colors are named for the apparent combination, such as yellowish-green, greenish-blue, and blue-violet. Normally only blue, green, yellow, red, white, and black appear as individual colors.[9] The reason for this is not definitely known but nevertheless all authorities agree that the phenomena of "pure hues" together with the "intermediate hues" are of considerable significance in color use. Their importance in cartography for showing interrelationship is obvious.

One of the peculiarities of hue sensitivity is the phenomenon of the stereoscopic effect of colors. Much of Peucker's theories of land form representation, and particularly his adaptive-spectral color scale, is based on the principle of advancing and retreating colors.[10] The advance of the warm colors and the retreat of the cool colors may be due partially to the sum of our experience in that warm colors are generally those of light while the cool are those of shadow.[11] The major cause of this sensation, long known in cartography, lies in the structure of the eye and the fact that light waves are refracted when they enter the eye. Rays are bent in inverse relation to their wave length, red rays being refracted least and blue most. The greater the refraction the farther away, relatively, the color seems. If the lens of the eye did not compensate for the differences in refraction, the warm hues would tend to focus behind the retina and the cool in front. The compensation of the lens draws the color nearer or pushes it back.[12] Although this phenomenon has been given great emphasis in cartography as previously noted in hypsometric and terrain shading, too much emphasis on any one aspect of color sensation, even though logical, is likely to produce undesirable results. The stereoscopic effect of color is not as strong a factor in the sum of sensations as, for example, are value differences, and therefore, although advancing and retreating colors can aid in creating an impression it is doubtful if they alone can produce a desired effect in cartography. This was well appreciated by Peucker and his colleagues.[13] On the other hand, this effect can be very detrimental if ignored as, for example, in the case of red boundaries or warm areas not "lying down on the map."

One of the most important effects of variation in hue sensitivity is that one associated with differences in visual acuity. It has long been known that monochromatic light is superior in defining power to mixed light and colors. In several tests results were obtained which demonstrated the superiority of monochromatic light and found yellow to be superior to other colors in defining power.[14] Value was held constant. It has also been demonstrated that the distinguishing of fine detail was markedly easier under monochromatic light even though value contrasts were somewhat lower. For distinguishing fine black lines, yellow was about 35 percent more efficient than blue, whereas there was somewhat less range between yellow and red.[15] Unfortunately other tests confuse hue and value. It would seem safe to conclude that nearly any hue which did not mix wave lengths too far apart would be reasonably efficient in defining power. Although chromatic aberration is of significance in visual acuity, value contrasts seem to be more important.[16]

It is not wise to single out any one component of design and place the greatest emphasis thereon since the possibilities of combination and the variety of utility are so great that in any one design some other component may dominate. Nevertheless, it is likely that, everything considered, no element assumes as great a role in design as does the interplay of light and dark (value differences). It is difficult to think of a map, which is anything like the ordinary, that does not have marked variations in value. Even the lettering exhibits value characteristics, not only as between the color of print and background as previously considered, but in the sense that a block of type or even a single word acts as a value area.[17] Value is one of the three sensations which the eye receives simultaneously, the others being hue and intensity. So far as is known no direct comparison among the three aspects of sensation has been made showing their relative significance, but there seems to be little doubt that value assumes greater importance in vision than either hue or intensity.

Value, as an element of vision, is of most significance when it enters into the application of contrast as an aspect of the effect of environment. There are, however, certain fundamental relationships in connection with the sensitivity of the eye to value not entirely based upon contrast effects.

The sensitivity of the eye to value differences is great over a wide range, but the sensitivity decreases as both extremely high and extremely low values are approached.[18] In other words, the eye requires greater differences in value near the black and near the white for a change to be noted. Of somewhat more significance in application, however, is the fact that with decreasing value the sensitivity diminishes more rapidly for colors of longer than of shorter wave length.[19] It is necessary, therefore, to plan for greater value differences in the lower scale for red than for blue. Both of the foregoing phenomena are of minor importance insofar as cartography is concerned since, generally speaking, value differences in cartography ought not to be so slight that such difference in sensitivity need be operative. Of far greater significance is the magnitude of the value change which is easily visible under normal illumination. Sargent points out that "without training, our eyes perceive easily about five degrees of value beginning with white and ending with black."[20] He goes on to say that with training one can recognize about ten graduations. Munsell chose ten with white and black at each end. Ostwald chose eight steps including black and white. Tone (brightness) densities in lithography and photography are usually limited to ten. It is apparent from both experimentation and from experience that the recognition of brightness values is relatively difficult and that six to eight steps from white to black is about the maximum for eyes with some training. To be sure, the specialist can recognize more, perhaps many times more, but for cartography six or eight values in addition to black and white seems to be the practical limit.

One of the basic laws of vision (and particularly significant in connection with value) is the Fechner or Weber-Fechner law. This law as applied to vision states that the sensation produced by an increase in stimulation bears a constant ratio to the total preceding stimulation. It is obvious then that a value visually midway between black and white will not be composed of half white and half black but rather of much less black than white. The application of this principle to cartographic technique is two-fold. Its first significance is in the employment of contrast, indicating that value contrast intervals must be based on sensation characteristics and not on simple arithmetic ratios.[21] Secondly, it is of fundamental importance in the field of statistical representation since value sensations will not be accurate in a scheme of simple areal distribution if the effects depend upon arithmetic relationships.[22]

Another principle of value sensitivity is that an extreme value, especially either white or black, tends to dominate a composition. Sargent points out that because of this characteristic of extreme values "large areas of both black and white . . . are seldom pleasing."[23] A great many maps, particularly special purpose maps, are today made with the basic value relationship being black water and white land. Since it appears that extreme value changes of large areas are likely to be fatiguing this practice is of questionable merit. The probable explanation of this phenomenon of fatigue is involved, but it may be connected with the observation that in practice black seems to be associated with the long wave end of the spectrum and white with the short end.[24] In the studies of color of print and background, black on yellow seems to be preferable from the point of view of visibility (although not for books and general reading) because it is less fatiguing than black and white contrasts.[25]

The sensitivity of the eye to changes in intensity has not been so well studied as has its sensitivity to value and hue. Studies summarized by Luckiesh seem to indicate that the eye is less sensitive to intensity than to either of the other primary sensations.[26] The just perceptible variations in intensity vary with hue and with value. The Munsell color tree clearly shows that the distinguishing of intensity variations is greatest in the middle values and falls off at each end of the scale. Studies made by Geissler apparently indicate that the intensity discrimination at equal values required a greater increment for the middle colors of the spectrum than for those at either end.[27]

Environment of color may, in general, be considered as having more significance in cartographic technique than intensity sensitivity, and as being of equal importance with hue and value sensitivity. Color environment assumes approximately the same degree of importance as contrast does in structural methodology. In a sense color always has an environment and the foregoing discussion of the sensitivity of the eye to hue, value and intensity, is therefore somewhat unrealistic in some of its aspects. For example value appearances will be markedly different depending upon environment; and the recognition of a true middle value between black and white is difficult, if not impossible, under contrasting environments.[28] Were it not for the effect of environment in modifying color sensations, the technique of map coloring would be relatively straightforward and would be capable of generally objective treatment. The general principles already considered are capable of measurement and many have been carried to a point where prediction of response is quite possible. The effects of environment, however, make it necessary to place in the background some of these principles and the actual selection of color must frequently be grounded on empirical bases. It is entirely possible, with sufficient testing and experience, that in the future even the facts of response to environmental conditions will be capable of exact determination.

The influence of environment on color sensation may be divided into two general categories, the first concerned with the effects of varying the physical conditions of viewing a color, and the second, and more important, dealing with the contrast effects of hue, value and intensity. This unusual importance is but another illustration of the importance in cartographic methodology of visual contrast.

A phenomenon which has not received much study is that associated with the changes in appearance of a colored area with variations in its size and position on the retina. Studies by MacDougal indicate that size of area has a definite effect on apparent intensity; and that the variations were greatest for the short wave end of the spectrum, and decreased steadily toward the long wave region.[29] It is apparent that when various colors are to be used to differentiate both large and small areas, changes in hue are more satisfactory than changes in intensity and brightness.

As already noted variations in illumination cause a shift in apparent hue of colors, and colors appear more saturated at low than at high illumination. Of much more significance however is the fact that the quality of light determines the appearance of colors, and hues arrived at under one kind of artificial light may not appear the same under some other kind of light or daylight. When combinations of colors are to be used on a map it is of importance to consider their appearance under the illumination in which they will be used.

Although the phenomenon of after image is interesting, and striking demonstrations are possible, it is doubtful if this effect is of more than passing interest in cartography. However, the phenomenon known as simultaneous contrast, dependent in part on after image effects, is of importance in planning a colored map. Simultaneous contrast is the sensation occurring when different colors, values, or intensities, lie adjacent to one another on the map. Complementary or near complementary colors when in juxtaposition enhance the intensity of one another.[30] As a consequence colors which are not particularly strong when viewed by themselves may be unduly brilliant when placed side by side. Doubtless this accounts in part for the unusual brilliance of some of the earlier political maps of the United States in which the states appeared in violently contrasting colors.

Perhaps the most common effects produced by simultaneous contrast are those involving value differences. Although the sensation so produced shows up in a number of ways it is basically a simple relationship. The greater the value contrast the greater the apparent value difference. This has been demonstrated many times.[31] The effects of simultaneous contrast are especially noticeable in maps employing gradations of value for progressive categories of data such as annual rainfall or population density. Simultaneous contrast often occurs in a most startling fashion when the values are placed next to one another in a legend. This gives rise to a phenomenon, known as induction, wherein the edge of an area of lighter value that is adjacent to an area of darker value will appear lighter than the other edge, especially if it is adjacent to an even lighter area.[32]

Consequently, the recognition of values is made difficult and great confusion will result if the effect of induction is also present on the map.

There are many other ways in which simultaneous contrast affects the character of colors most of which are of lesser importance in cartography. Hue, value, and intensity singly, or in any combination, can be made to appear differently according to background.[33] The effects of simultaneous contrast, including induction, arc largely eliminated if the contrasting areas are separated or if the areas are outlined in black.[34]

Another significant phenomenon caused by simultaneous contrast is the apparent increase in size of areas or lines as they are increased in brilliance.[35] Although there have been a number of tests of the effect of hue on the apparent size of objects,[36] none of them seems to have gone far enough to derive any mathematical relationships. They all show that light objects seem larger and dark ones smaller. Hue apparently has little effect.

It is a common practice in cartography to justify the use of color on maps by the logic of association with the thing mapped. Thus the ocean is blue; lowlands (European!) are green; red is for high temperatures; blue, purple, and white are cold; soil is brown; and many others. The superlogical mind enjoys the sport of color association, and the extremes to which it can be carried are well illustrated by the following quotation from Eckert:

> An interesting and logical use of colors is shown by the maps belonging to the field of medicine and which were produced for the first time at the beginning of this century by Rossle. On a panoramic map of Germany he presented the spread and frequency of occurrence of dangerous illnesses. For tuberculosis he chose blue; the diseased lung looks gray-blue. Blue, for cartographical and statistical representation of tuberculosis, has already reached international meaning in the world of medicine. For the spread of child diseases, such as scarlet fever and measles, the red color was chosen; typhus, characterized by its dark brown bowel movement, is painted dark brown; and cholera is shown with yellow-brown colors.[37]

Figure 3. Induction. Each gray patch is of uniform value, but the adjacent value modifies its appearance. (From an Eastman gray scale.)

The use of the spectral hues in the order in which they appear is strongly entrenched in cartography, not only for hypsometric shadings, but for other gradations of data as well.[38] Eckert, however, was quite open-minded and pointed out that perhaps other systems might be preferable, for he refers to the possibility of using the Ostwald "color circle" and states: "Up till now [1938] no attempts have been made to use the color circle in cartography" . . . "I have great faith in it."[39]

There seems to be little reason to believe that the cartographic technique will not ultimately leave behind its dependence on the early physical and physiological researches in color and advance toward the utilization of the research of the psychologists and others investigating the effects of color as a sensation of the human eye. This by no means requires the discard of all color conventions. Red and blue stand at the top of the list of color preferences. Traditionally red is warm and blue is cool, although a study by Morgansen and English seemed to show opposite results.[40] The traditional significance and appropriateness of colors is a subject about which much has been written and many of the theories are somewhat in conflict.[41] A large number of psychological studies have been made in the field of the significance of color and more especially in connection with color preference.[42] Although the results of these tests are not conclusive they tend to suggest safe limits for the methodology of color choice.

In general the evidence and logic seem preponderantly in favor of basing the utilization of color in designing maps on the fact that color as it affects the human eye is primarily a sensation, the ramifications of which bear little logical relation to the arrangement of hues and value in the spectrum or to other strained associations. Out of the extreme complexities of the sensations of color, the variations of value contrast appear to be most effective in presentation. The other aspects of color, although of somewhat less significance generally, may be of extreme importance in special cases and may contribute significantly to the development of clear, unequivocal, and legible cartography.

NOTES

1. Matthew Luckiesh, *Color and its Applications* (New York, 1927), pp. 124–126.
2. The International Printing Ink Corp. Research Laboratories, "Color as Light," Monograph No. 2 (New York, 1935), pp. 13–20.
3. Birren, *The Story of Color*, p. 259.
4. Ibid., p. 309.
5. Matthew Luckiesh, *The Language of Color* (New York, 1918), p. 161.
6. Luckiesh, *Color and Its Applications*, pp. 208–211.
7. Encyclopaedia Britannica, *World Atlas*, pp. 3F–3Q.
8. Cf. Walter Sargent, *The Enjoyment and Use of Color* (New York, 1923), p. 44.
9. Birren, *The Story of Color*, p. 257.
10. See Eckert, *Die Kartenwissenschaft*, I, pp. 625–639, especially 631–634.
11. Sargent, p. 55.
12. Cf. Birren, *The Story of Color*, pp. 280–281.
13. Eckert, *Die Kartenwissenschaft*, I, pp. 625 ff.
14. Luckiesh, *Color and Its Applications*, pp. 129–137.
15. Ibid.
16. Cf. Miles A. Tinker, "The Effect of Color in Visual Apprehension and Perception," *Genetic Psych.*, Monog. 11 (1931), pp. 61–136.
17. Cf. Eugene De Lopatecki, "Type Color," *Seventh Annual Advertising and Publishing Production Yearbook* (New York, 1941), pp. 330–331. "Color of a type means the degree of blackness which a block of type registers against a white background. Good layout calls for specific weight for each type of block."
18. Luckiesh, *Color and Its Applications*, p. 720.
19. Ibid.
20. Sargent, p. 62.
21. Cf. Lithographic Technical Foundation, "The Relation between Dot Area, Dot Density, and Tone Value for Half-tone Images," *Tech. Bull. No. 5* (New York, 1945).
22. Cf. Arthur H. Robinson, "A Method for Producing Shaded Relief from Areal Slope Data," *Annals Assn. of Amer. Geog.*, XXXVI (1946), 248–252. See also George Babcock Cressey, *China's Geographic Foundations* (New York, 1934), Figs. 46, 47, 48, 49, on which "the area covered by each dot is roughly equal to the actual area which it represents."
23. Sargent, pp. 70–71.

24. Ibid., pp. 72–74.
25. Birren, *The Story of Color*, p. 309.
26. Luckiesh, *Color and Its Applications*, pp. 127–129.
27. Ibid.
28. Ibid., p. 175.
29. Ibid., pp. 163–164.
30. Ibid., p. 173; also Sargent, p. 119.
31. E.g., Luckiesh, p. 174; and The International Printing Ink Corp., Monograph II, p. 11.
32. Luckiesh, *Color and Its Applications* p. 175.
33. Good examples are given in The International Printing Ink Corp., Monograph III, p. 11.
34. Luckiesh, *Color and Its Applications*, p. 176.
35. Ibid., p. 179.
36. See Albert R. Chandler and Edward N. Barnhart, *A Bibliography of Psychological and Experimental Aesthetics* (Berkeley, Univ. of California Press, 1938), pp. 30–35.
37. Eckert-Greifendorff, *Kartographie*, pp. 41–42.
38. Ibid., pp. 42–43.
39. Ibid.,, p. 43.
40. M. F. Morgansen and H. B. English, "The Apparent Warmth of Colors," *Am. Jour. of Psych.*, XXXVII (1926), 427–428.
41. See M. Luckiesh, *The Language of Color*; and Birren, *The Story of Color*, Appendixes A and B, 323 ff.
42. Chandler and Barnhart, pp. 37–45, 170.

A SELECTED BIBLIOGRAPHY

A selected bibliography

Adams, Cyrus A. "Maps and Map Making." *Bull. Of the Amer. Geog. Soc.*, XLIV (1912), 194–201.

Alexander, John W., and Zahorchak, George A. "Population Density Maps of the United States: Techniques and Patterns," *Geog. Rev.*, XXXIII (1943), 457–466.

American Society for Professional Geographers. Committee on Cartography. "Cartography for Geographers," *The Professional Geographer*, IV (1946), 10–12.

Becker, F. "Die Kunst in der Kartographie," *Geographische Zeitschrift*, XVI (1910), 473–490.

Birdseye, C. H. "Topographic Instructions," *Geological Survey Bull. 788*. Washington, Dept. of the Interior, 1928.

Birren, Faber. "A Chart to Illustrate the Fundamental Discrepancies of Color Study in Physics, Art, and Psychology," *Psych. Rev.*, XXXVII (1930), 271–273.

_________. *Color Dimensions*. Chicago, 1934.

_________. *The Story of Color*. Westport, Conn., 1941.

Bludau, Alois. "Uber die Wahl der Projectionen für die Landkarten der Hand und Schulatlanten," *Geographische Zeitschrift*, I (1895), 497–516.

Boggs, S. W. "Cartohypnosis," *Scientific Monthly*, LXIV (1947), 469–476.

Bragg, William. *The Universe of Light*. New York, 1933.

Burtt, Harold E. *Psychology of Advertising*. Boston, 1938.

_________ and Basch, C. "Legibility of Bodoni, Baskerville, Roman, and Cheltenham Typefaces," *Jour. of Appl. Psych.*, VII (1923), 237–245.

Chamberlin, Wellman. *The Round Earth on Flat Paper*. Washington, 1947.

Chandler, Albert R., and Barnhart, Edward N. *A Bibliography of Psychological and Experimental Aesthetics*. Berkeley, Univ. of Calif. Press, 1938.

Clodd, Edward. *The Story of the Alphabet*. New York, D. Appleton & Co., 1901. 204 p. illus.

Dashiell, John Frederick. *Fundamentals of Objective Psychology*. Cambridge, Mass., 1928.

Davis, W. M. "The Principles of Geographical Description," *Annals Assn. of Amer. Geog.*, V (1915), 61–105.

_________. "The Progress of Geography in the United States," *Annals Assn. of Amer. Geog.*, XIV (1924), 159–215.

De Lopatecki, Eugene. *Advertising Layout and Typography*. New York, 1935.

_________. "Type Color," *Seventh Annual Advertising and Publishing Yearbook*. New York, 1941.

_________. *Typographer's Desk Manual*. New York, 1937.

Durham, Henry W. "A Plea for Real Maps," *Engineering News Record*, June (1940), 44–45.

Eckert, Max. *Die Kartenwissenschaft*, 2 Vols. Berlin and Leipzig, 1921, 1925.

_________. "On the Nature of Maps and Map Logic," trans. by W. [L. G.] Joerg, *Bull. of the Amer. Geog. Soc.*, XL (1908), 344–351.

Eckert-Greifendorff, Max. *Kartographie*. Berlin, 1939.

Eldredge, A. G., Abrams, A. W., Jansen, W., and Shyrock, C. M. "Maps and Map Standards," *The Teaching of Geography*. 32nd Yearbook of the Nat. Soc. for the Study of Education. Bloomington, Ill., 1933. Chap. 25.

Evans, Ralph M. *An Introduction to Color*. New York, 1948.

Fead, M. I. "Notes on the Development of the Cartographic Representation of Cities," *Geog. Rev.*, XXIII (1933), 441–456.

Finch, V. C. "A Greater Appreciation of Maps," *The Business Education World*, XVIII, 4 (1937), 1–5.

Fisher, Irving, and Miller, O. M. World Maps and Globes. New York, 1944.

Frazier, J. L. *Modern Type Display*. Chicago, 1920.

Genthe, M. K. "Notes on the History of Gotha Cartography," *Bull. of the Amer. Geog. Soc.*, XLV (1913), 33–38.

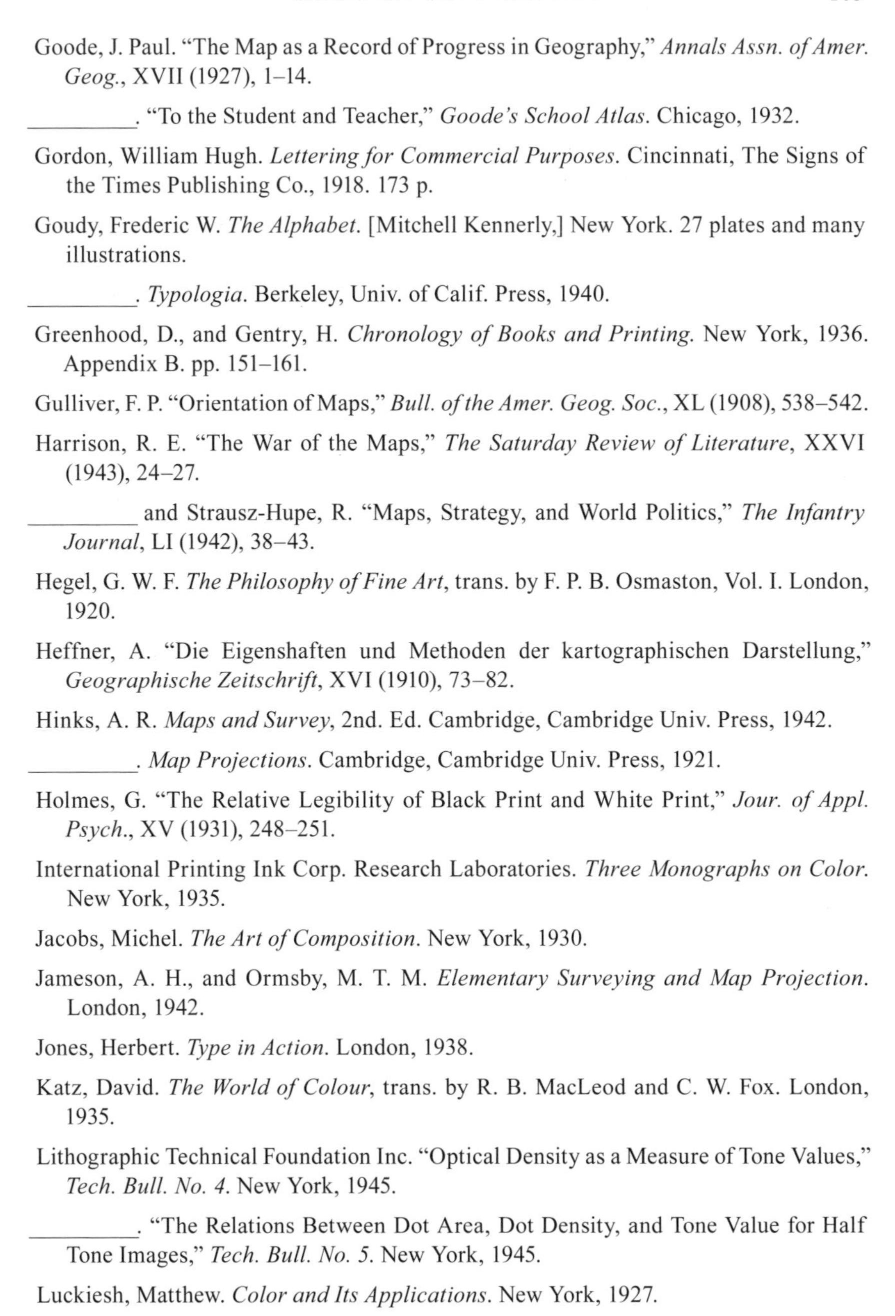

Goode, J. Paul. "The Map as a Record of Progress in Geography," *Annals Assn. of Amer. Geog.*, XVII (1927), 1–14.

_________. "To the Student and Teacher," *Goode's School Atlas*. Chicago, 1932.

Gordon, William Hugh. *Lettering for Commercial Purposes*. Cincinnati, The Signs of the Times Publishing Co., 1918. 173 p.

Goudy, Frederic W. *The Alphabet*. [Mitchell Kennerly,] New York. 27 plates and many illustrations.

_________. *Typologia*. Berkeley, Univ. of Calif. Press, 1940.

Greenhood, D., and Gentry, H. *Chronology of Books and Printing*. New York, 1936. Appendix B. pp. 151–161.

Gulliver, F. P. "Orientation of Maps," *Bull. of the Amer. Geog. Soc.*, XL (1908), 538–542.

Harrison, R. E. "The War of the Maps," *The Saturday Review of Literature*, XXVI (1943), 24–27.

_________ and Strausz-Hupe, R. "Maps, Strategy, and World Politics," *The Infantry Journal*, LI (1942), 38–43.

Hegel, G. W. F. *The Philosophy of Fine Art*, trans. by F. P. B. Osmaston, Vol. I. London, 1920.

Heffner, A. "Die Eigenshaften und Methoden der kartographischen Darstellung," *Geographische Zeitschrift*, XVI (1910), 73–82.

Hinks, A. R. *Maps and Survey*, 2nd. Ed. Cambridge, Cambridge Univ. Press, 1942.

_________. *Map Projections*. Cambridge, Cambridge Univ. Press, 1921.

Holmes, G. "The Relative Legibility of Black Print and White Print," *Jour. of Appl. Psych.*, XV (1931), 248–251.

International Printing Ink Corp. Research Laboratories. *Three Monographs on Color*. New York, 1935.

Jacobs, Michel. *The Art of Composition*. New York, 1930.

Jameson, A. H., and Ormsby, M. T. M. *Elementary Surveying and Map Projection*. London, 1942.

Jones, Herbert. *Type in Action*. London, 1938.

Katz, David. *The World of Colour*, trans. by R. B. MacLeod and C. W. Fox. London, 1935.

Lithographic Technical Foundation Inc. "Optical Density as a Measure of Tone Values," *Tech. Bull. No. 4*. New York, 1945.

_________. "The Relations Between Dot Area, Dot Density, and Tone Value for Half Tone Images," *Tech. Bull. No. 5*. New York, 1945.

Luckiesh, Matthew. *Color and Its Applications*. New York, 1927.

_________. *The Language of Color*. New York, 1918.

_________. *Light and Color in Advertising and Merchandising.* New York, 1923.

_________ and Moss, F. K. "The Quantitative Relationship between Visibility and Type Size," *Jour. of the Franklin Inst.*, CCXXVII (1939), 87–97.

_________. "The Visibility of Various Typefaces," *Jour. of the Franklin Inst.*, CCXXIII (1937), 76–82.

Luckiesh, Matthew. "Visibility and Readability of Print on White and Tinted Papers," *Sight Saving Rev.*, VIII (1938), 123–124.

Lyons, H. G. "Relief in Cartography," *Geog. Jour.*, XLIII (1914), 233–248, 395–398.

MacDonald, D. J., and Hart, G. *Survey of Lithography.* New York, 1945.

Marschner, F. J. "Maps and a Mapping Program for the United States," *Annals Assn. of Amer. Geog.*, XXXIII (1943), 199–219.

_________. "Structural Properties of Medium and Small Scale Maps," *Annals Assn. of Amer. Geog.*, XXXIV (1944), 1–46.

Miller, O. M. "An Experimental Air Navigation Map," *Geog. Rev.*, XXIII (1933), 48–60.

Miyake, M. F., Dunlap, J. W., and Cureton, E. E. "The Comparative Legibility of Black and Colored Numbers on Colored and Black Backgrounds," *Jour. of Gen. Psych.*, III (1930), 340–343.

Monsen, M. T. "The Importance of Type in Lithography," *The Lithographers Manual.* New York, 1940. pp. 72–73a.

Morgansen, M. E., and English, H. B. "The Apparent Warmth of Colors," *Amer. Jour. of Psych.*, XXXVII (1926), 427–428.

Munsell, Albert H. *A Color Notation.* Boston, 1916.

Ostwald, Wilhelm. *Colour Science*, 2 Vols., trans. by J. Scott Taylor. London, 1931–1933.

Paterson, D. G., and Tinker, M. A. *How to Make Type Readable.* New York, 1940.

Poffenberger, A. T. *Psychology in Advertising.* New York, 1932.

_________ and Franken, R. B. "A Study of the Appropriateness of Typefaces," *Jour. of Appl. Psych.*, VII (1923), 312–329.

Preston, K., Schwankl, H. P., and Tinker, M. A. "The Effect of Variations of Color of Print and Background on Legibility," *Jour. of Gen. Psych.*, VI (1932), 459–461.

Preuss, Wolfgang. "Zur Darstellung von Bevölkerungsverteilung und Volksdichte," *Mit. d. Ver. d. Geographen a. d. Univ. d. Leipzig*, Heft. 14/15 (1936), 67–81.

Raisz, Erwin. "Draw Your Own Blackboard Maps," *Jour. of Geog.*, XLI (1942), 236.

_________. *General Cartography.* New York, 1938.

_________. "Map Projections and the Global War," *The Teaching Scientist*, II (1946), 33–39.

_________. "The Mapping of the Earth, Past, Present, and Future," *Geog. Jour.*, XLVIII (1916), 331–346.

Ridgeway, John L. *Scientific Illustration*. Stanford, Stanford Univ. Press, 1938.

Roethlein, B. E. "The Relative Legibility of Different Faces of Printing Types," *Publ. of the Clark Univ. Lib.*, III (1912), 1–41.

Ross, Denman W. *A Theory of Pure Design*. Boston and New York, 1907.

Sargent, Walter. *The Enjoyment and Use of Color*. New York, 1923.

Soffner, Hans. "War on the Visual Front; Use of Maps, Charts and Diagrams for Purposes of Propaganda," *Amer. Scholar*, XI (1942), 465–476.

Southall, James P. C. *Introduction to Physiological Optics*. London, Oxford Univ. Press, 1937.

Speier, Hans. "Magic Geography," *Social Research*, VIII (1941), 310–330.

Spykman, Nicholas. "Mapping the World," *The Geography of Peace*. New York, 1944. Chap. 2.

Stanton, F. N., and Burtt, H. E. "The Influence of Surface and Tint of Paper on Speed of Reading," *Jour. of Appl. Psych.*, XIX (1935), 683–693.

Steers, J. A. *The Study of Map Projections*, 7th Ed. London, 1949.

Stewart, J. Q. "The Use and Abuse of Map Projections," *Geog. Rev.*, XXXIII (1943), 589–604.

Sumner, F. C. "Influence of Color on Legibility of Copy," *Jour. of Appl. Psych.*, XVI (1932), 201–204.

Taylor, C. D. "The Relative Legibility of Black and White Print," *Jour. of Educational Psych.*, XXV (1934), 561–578.

Taylor, Canon Isaac. The Alphabet, Its Origin and Development, 2 Vols. London, 1899.

Tinker, Miles A. "The Effect of Color in Visual Apprehension and Perception," *Genetic Psych.*, Monograph XI (1931), 61–63.

_________ and Paterson, D. G. "Variations in Color of Print and Background," *Jour. of Appl. Psych.*, XV (1931), 471–479.

Tissot, Auguste. *Memoire sur la Représentation des Surfaces et les Projections des Cartes Geographiques*. Paris, 1881.

Updike, Daniel Berkeley. *Printing Types, Their History, Forms, and Uses; A Study in Survivals*, 2 Vols. Cambridge, Harvard Univ. Press, 1922. 275; 308 p. illus.

Wagner, Hermann. "Zur Geschichte der Gothaer Kartographie," *Petermann's Mit.*, LVIII, 1 (1912), 76–79.

Wirsing, Giselher. *The War in Maps, German Library of Information*. New York, 1941.

Withycombe, J. G. "Lettering on Maps," *Geog. Jour.*, LXXIII (1929), 429–446.

Wright, John K. "Map Makers Are Human. Comments on the Subjective in Maps," *Geog. Rev.*, XXXII (1942), 527–544.

_________. "The World in Maps," *Geog. Rev.*, XXX (1940), 1–18.

Related titles from ESRI Press

Cartographic Relief Presentation

ISBN: 978-1-58948-026-1

Originally published in 1965, *Cartographic Relief Presentation* provides guidelines for properly rendering terrain in maps of all types and scales. This book is an example of the art of combining cartography with intellect and graphics when solving map design problems.

Designed Maps: A Sourcebook for GIS Users

ISBN: 978-1-58948-160-2

This companion to the highly successful *Designing Better Maps* offers a graphics-intensive presentation of published maps, providing cartographic examples that GIS users can adapt for their own needs. Visual hierarchies, and the purpose and audience of each map are considered, drawing a clear connection between intent and design.

Designing Better Maps: A Guide for GIS Users

ISBN: 978-1-58948-089-6

Designing Better Maps demystifies the basics of good cartography, walking readers through layout design, scales, north arrows, projections, color selection, font choices, and symbol placement. Recognizing the need for integration with other publishing and design programs, the text also covers various export options, all of which lead to the creation of publication-worthy maps.

Semiology of Graphics: Diagrams, Networks, Maps

ISBN: 978-1-58948-261-6

Originally published in 1967, *Semiology of Graphics* established the theoretical framework for the study of information visualization. Today, Jacques Bertin's classic tome is reawakened by the limitless possibilities presented by modern technology.

ESRI Press publishes books about the science, application, and technology of GIS. Ask for these titles at your local bookstore or order by calling 1-800-447-9778. You can also read book descriptions, read reviews, and shop online at www.esri.com/esripress. Outside the United States, contact your local ESRI distributor.